An Introduction to SAS® University Edition

Ron Cody

support.sas.com/bookstore

The correct bibliographic citation for this manual is as follows: Cody, Ron. 2015. *An Introduction to SAS® University Edition.* Cary, NC: SAS Institute Inc.

An Introduction to SAS® University Edition

ISBN 978-1-62959-770-6 (Hard copy)
ISBN 978-1-62960-007-9 (EPUB)
ISBN 978-1-62960-008-6 (MOBI)
ISBN 978-1-62960-010-9 (PDF)

SAS Institute Inc., SAS Campus Drive, Cary, NC 27513-2414

February 2018

Contents

List of Programs

Chapter 7

Chapter 8

Chapter 9

Chapter 10

Chapter 11

Chapter 12

Chapter 13

Chapter 14

Chapter 15

Chapter 16

Chapter 17

Chapter 18

Chapter 19

Chapter 20

About This Book

Is This Book for You?

Whether you are learning SAS on your own or as a student in a class that includes SAS, this book should be your constant companion. For teachers, consider using *Introduction to SAS University Edition* for any course that includes SAS programming, especially courses that use the SAS University Edition.

Although extensive online information describes how to use the SAS University Edition, readers will appreciate a single book that has everything they need in one place. The book's programming section relies heavily on simple examples. By talking to many students and SAS users, the author has discovered that most programmers prefer this method of learning. They will say, "Don't show me the syntax—show me an example!" Again, because all of the programs and data sets are available in a free download, you can run all the examples and then use those examples as a starting point for your own programs.

Scope of This Book

The first part of the book describes the process of installing and configuring the SAS University Edition and using its built-in tasks. These tasks include a tool for importing data from Excel workbooks and CSV files along with many of the other popular file types such as Access and SPSS. Part 1 shows you how to create reports, summarize data, compute frequencies, and generate charts and graphs.

Part 2 teaches you how to use SAS as a programming language. This section of the book is also helpful to SAS users who employ other interfaces such as SAS Enterprise Guide or the SAS Display manager.

About the Examples

Software Used to Develop the Book's Content

The SAS University Edition was used to develop the content for this book.

Example Code and Data

You can access the example code and data for this book by linking to its author page at http://support.sas.com/publishing/authors. Select the name of the author. Then, look for the cover thumbnail of this book, and select **Example Code and Data** to display the SAS programs that are included in this book.

For an alphabetical listing of all books for which example code and data are available, see http://support.sas.com/bookcode. Select a title to display the book's example code.

If you are unable to access the code through the web site, send email to saspress@sas.com.

Exercise Solutions

The answers to odd-numbered problems can be found at the back of the book in the appendix. For students, the answers to the even-numbered problems are available to teachers or by special request.

Additional Help

Although this book illustrates many analyses regularly performed in businesses across industries, questions specific to your aims and issues may arise. To fully support you, SAS Institute and SAS Press offer you the following help resources:

- For questions about topics covered in this book, contact the author through SAS Press:
  - Send questions by email to saspress@sas.com; include the book title in your correspondence.
  - Submit feedback on the author's page at http://support.sas.com/author_feedback.
- For questions about topics in or beyond the scope of this book, post queries to the relevant SAS Support Communities at https://communities.sas.com/welcome.
- SAS Institute maintains a comprehensive web site with up-to-date information. One page that is particularly useful to both the novice and the seasoned SAS user is its Knowledge Base. Search for relevant notes in the "Samples and SAS Notes" section of the Knowledge Base at http://support.sas.com/resources.
- Registered SAS users or their organizations can access SAS Customer Support at http://support.sas.com. Here you can pose specific questions to SAS Customer Support; under **Support**, click **Submit a Problem**. You will need to provide an email address to which replies can be sent, identify your organization, and provide a customer site number or license information. This information can be found in your SAS logs.

Keep in Touch

We look forward to hearing from you. We invite questions, comments, and concerns. If you want to contact us about a specific book, please include the book title in your correspondence.

Contact the Author through SAS Press

- By email: saspress@sas.com
- On the web: http://support.sas.com/author_feedback

Purchase SAS Books

For a complete list of books available through SAS, visit sas.com/store/books.

- Phone: 1-800-727-0025
- Email: sasbook@sas.com

Subscribe to the SAS Training and Book Report

Receive up-to-date information about SAS training, certification, and publications through email by subscribing to the SAS Training & Book Report monthly eNewsletter. Read the archives and subscribe today at http://support.sas.com/community/newsletters/training!

Publish with SAS

SAS is recruiting authors! Are you interested in writing a book? Visit http://support.sas.com/saspress for more information.

About The Author

Ron Cody, EdD, a retired professor from the Rutgers Robert Wood Johnson Medical School now works as a private consultant and a national instructor for SAS Institute Inc. A SAS user since 1977, Ron's extensive knowledge and innovative style have made him a popular presenter at local, regional, and national SAS conferences. He has authored or co-authored numerous books, such as *Learning SAS by Example: A Programmer's Guide*; *SAS Statistics by Example, Applied Statistics and the SAS Programming Language, Fifth Edition*; *The SAS Workbook*; *The SAS Workbook Solutions*; *Cody's Data Cleaning Techniques Using SAS, Second Edition*; *Longitudinal Data and SAS: A Programmer's Guide*; *SAS Functions by Example, Second Edition*, and *Cody's Collection of Popular Programming Tasks and How to Tackle Them*, and *Test Scoring and Analysis Using SAS,* as well as countless articles in medical and scientific journals.

Learn more about this author by visiting his author page at http://support.sas.com/cody. There you can download free book excerpts, access example code and data, read the latest reviews, get updates, and more.

Acknowledgments

It takes a village—well not quite, but a great team of people helped bring this book to fruition. Sian Roberts did double duty, acting both as the acquisitions editor and developmental editor. Sian helped produce my last book on test scoring, and it was a pleasure to work with her again.

Next, I wish to thank the technical reviewers: Paul Grant, Amy Peters, Jennifer Tamburro, Kathy Passarella, Marie Dexter, Mark Jordan, and Catherine Gihlstorf. The technical review is one of the most important steps in developing a book such as this. These reviewers read every line of text and checked every line of code.

I am always amazed at how well the copy editors perform a very difficult job. Kudos to Kathy Restivo and Mary Beth Steinbach for fixing my typos and suggesting alternate wording. Their job was especially difficult with this book because there was pressure to get the book into the hands of all of the folks who downloaded the SAS University Edition as quickly as possible.

The final step in bringing this book to market was provided by Monica McClain, the technical publishing specialist. This is a demanding job that requires attention to detail. Thank you, Monica.

Last, but certainly not least, I want to thank Jan Cody and Robert Harris. Thank you, Jan (my wife), for the great photo of me included on the back cover and in the book. Robert created the cover design, and not only did he create a cover that grabs your attention, he created two other cover designs for me to choose from. They were all great and I had a difficult time choosing the one you see in front of you.

Ron Cody
Fall 2015

Part 1: Getting Acquainted with the SAS Studio Environment

Part I shows you how to perform basic tasks, such as producing a report, summarizing data, producing charts and graphs, and using the SAS Studio built-in tasks.

Chapter 1: Introduction to the SAS University Edition

Introduction: An Overview of SAS and SAS University Edition

SAS is many things: A data analysis tool, a programming language, a statistical package, a tool for business intelligence, and more. Until recently, you could get access to SAS by paying a license fee or, if you were a college student, you could use a free version of SAS called "SAS onDemand for Academics."

The really big news is that anyone can now obtain SAS for FREE! It's called SAS University Edition and anyone can download a copy and run it on just about any computer. So, what's the catch? SAS University Edition is for non-commercial use only.

On many college campuses, students taking statistics courses or any course that needs a powerful analytic tool could access a computer language called R for free. Since free is better than not free, these institutions sometimes choose to use R instead of SAS. That is fine, except that when these students graduate, they find that in the corporate world, SAS is by far the preferred package for powerful statistical analysis, data manipulation, and reporting. By offering a free version of SAS, users now have a choice between SAS or R and SAS Institute is hoping that the majority of users will choose SAS.

SAS University Edition uses SAS Studio as the interface. SAS Studio provides an environment that includes a point-and-click facility for performing many common tasks, such as producing reports, graphs, data summaries, and statistical tests. For those who either enjoy programming or have more complicated tasks, SAS Studio also allows you to write and run your own programs.

Obtaining Your Free Copy of SAS

To obtain your free copy of SAS University Edition, use the following URL:

http://www.sas.com/en_us/software/university-edition.html

This will bring up the screen shown here:

Figure 1.1: Obtaining Your Free Copy of SAS University Edition

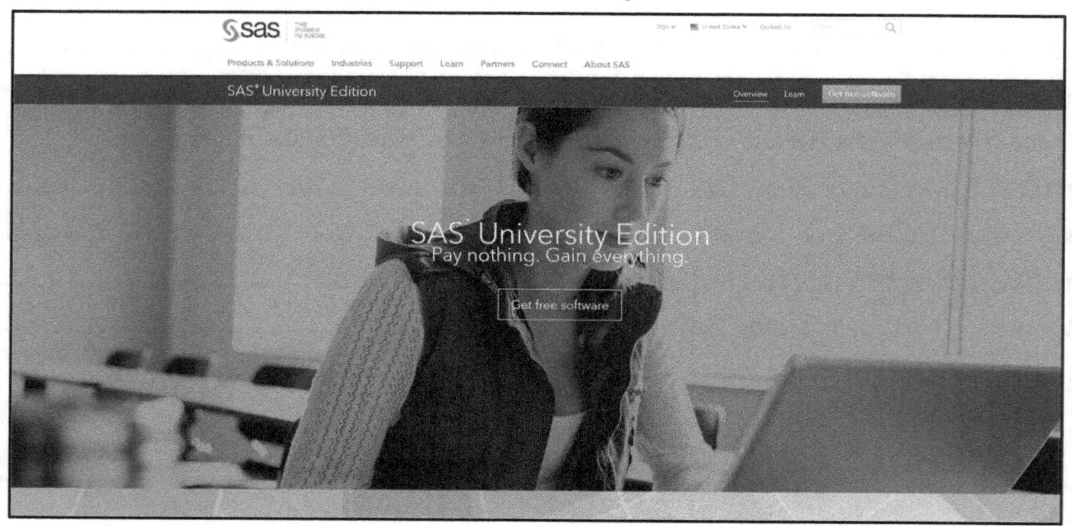

SAS University Edition runs in a virtual environment on any computer that can run either Oracle VirtualBox, VMware Workstation Player, or VMware Fusion (for Apple computers). Requirements for running the SAS University Edition are displayed when you request a download of SAS University Edition.

The screenshots in the remaining portions of this chapter demonstrate running SAS University Edition using VirtualBox. If you choose to use VMware Workstation Player or Vmware Fusion, just follow the on-screen directions and everything should go smoothly.

Click on the "Get the Software" button at the top right then scroll down to select your operating system (Figure 1.2).

Figure 1.2: Choose Your Operating System

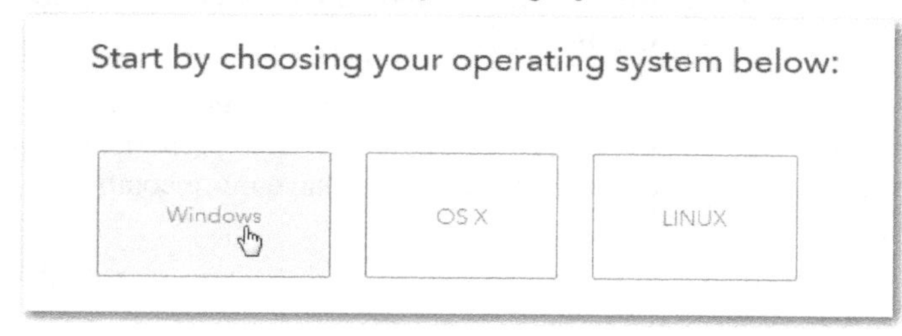

If you don't already have virtualization software on your computer, the SAS download page provides links to obtain virtualization software for your computer and operating system. You will see a list of requirements to run the visualization software. If you have an older operating system, you may still be able to run an older version of VirtualBox or other virtualization software.

If you are not an expert in any of these virtual computer packages, be sure to download the PDF and **read it.** Also, watch the videos that are included in each step; they are excellent.

Once you have selected your operating system, scroll down until you see the following:

Figure 1.3: Steps to Download SAS University Edition

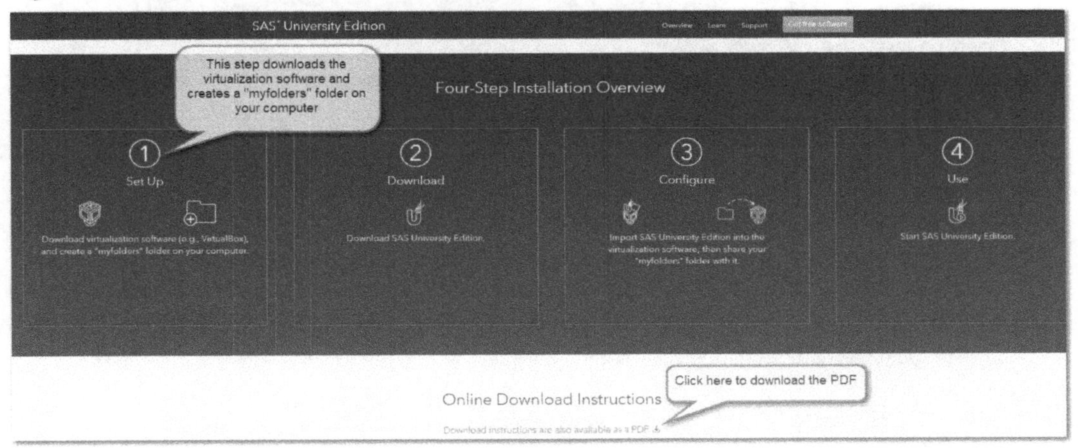

Step 1: Download VirtualBox and Create the "myfolders" Folder on your Computer

At this point you can view a video or continue to scroll down to download the visualization software. You should see the following:

Figure 1.4: See the Video, Download the Virtualization Software, or Create a Shared Folder

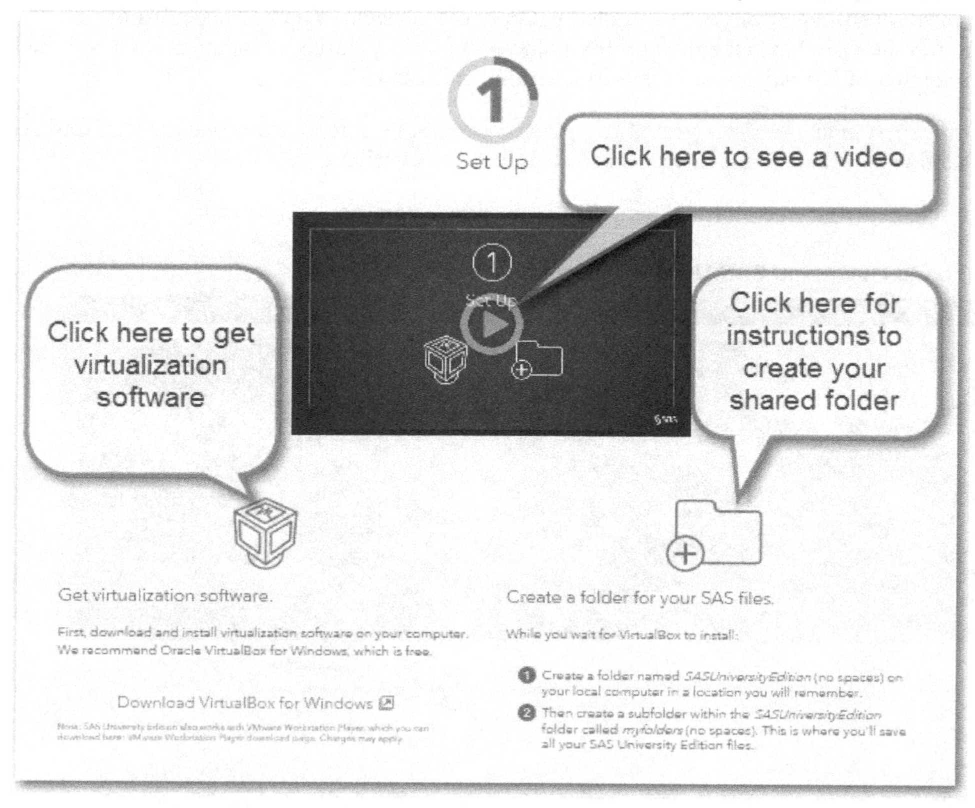

After you view the video (recommended), proceed to download the virtualization software.

Figure 1.5: Get the Virtualization Software

While you are waiting for the virtualization software to load, follow the instructions to set up a shared folder on your computer, as shown in the screen shot below:

Figure 1.6: Create a Shared Folder for your SAS Files

Step 2: Download SAS University Edition

Once you have downloaded VirtualBox (or other virtualization software) and created your folder, you are ready to download the SAS University Edition virtual application (vApp) and store it in your downloads directory.

Figure 1.7: Downloading the SAS University Edition

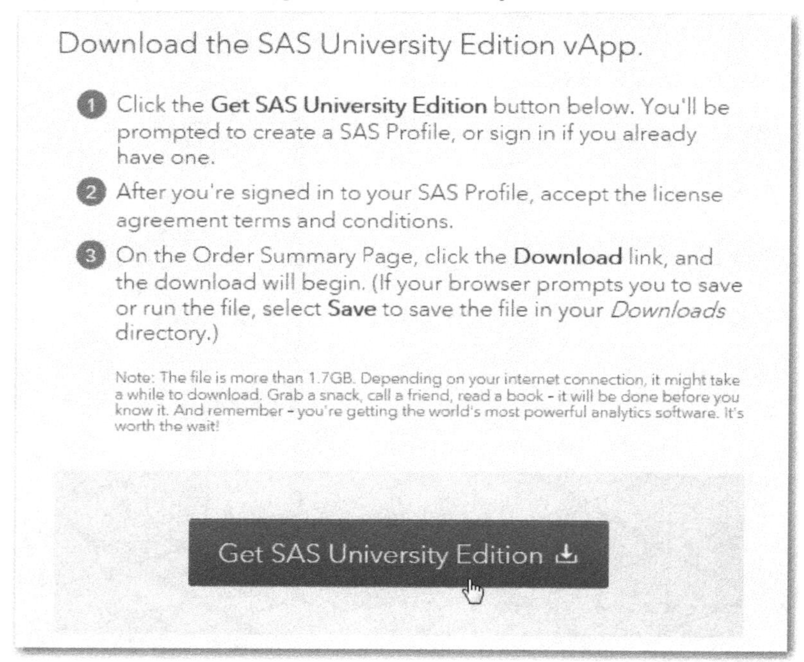

Step 3: Configure VirtualBox and Set up the Shared Folder

Open VirtualBox and select the option to open a virtual machine:

Figure 1.8: Importing SAS University Edition into VirtualBox

In an earlier step, you set up a folder called SASUniversityEdition\myfolders on your computer. Now you need set up VirtualBox so it can access this folder. The next screen shot demonstrates how this is accomplished:

Figure 1.9: Setting up a Shared Folder in VirtualBox

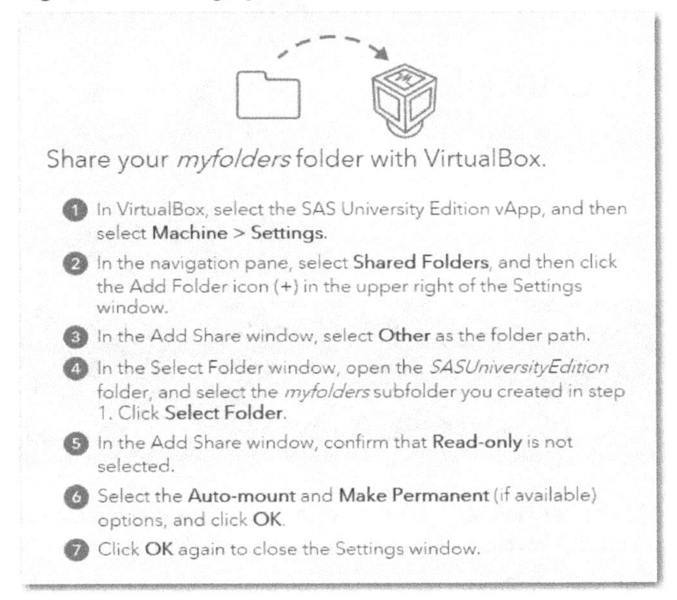

Share your *myfolders* folder with VirtualBox.
1 In VirtualBox, select the SAS University Edition vApp, and then select **Machine > Settings**.
2 In the navigation pane, select **Shared Folders**, and then click the Add Folder icon (+) in the upper right of the Settings window.
3 In the Add Share window, select **Other** as the folder path.
4 In the Select Folder window, open the *SASUniversityEdition* folder, and select the *myfolders* subfolder you created in step 1. Click **Select Folder**.
5 In the Add Share window, confirm that **Read-only** is not selected.
6 Select the **Auto-mount** and **Make Permanent** (if available) options, and click **OK**.
7 Click **OK** again to close the Settings window.

Step 4: Use - Start University Edition

All the hard work is behind you. From now on, you can simply click on a bookmark to run SAS University Edition (after you start or restart your virtual machine). The way you run SAS University Edition in VirtualBox is shown next (the URL to access SAS is different if you use other virtualization software). Remember to leave VirtualBox (or other virtualization software) running.

Figure 1.10: Running SAS University Edition

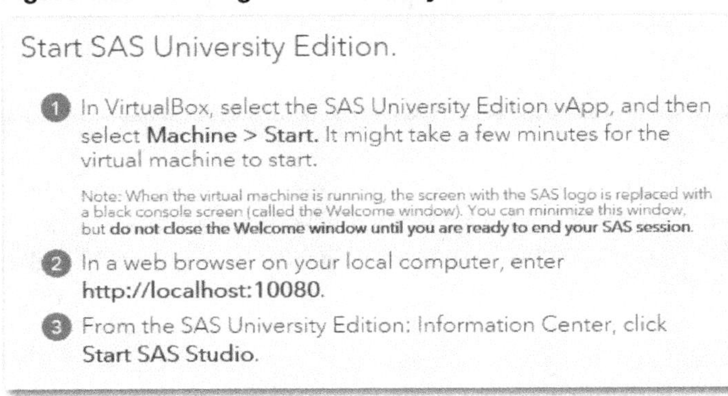

Start SAS University Edition.
1 In VirtualBox, select the SAS University Edition vApp, and then select **Machine > Start.** It might take a few minutes for the virtual machine to start.
Note: When the virtual machine is running, the screen with the SAS logo is replaced with a black console screen (called the Welcome window). You can minimize this window, but **do not close the Welcome window until you are ready to end your SAS session**.
2 In a web browser on your local computer, enter **http://localhost:10080**.
3 From the SAS University Edition: Information Center, click **Start SAS Studio**.

Here is a screen shot of what you see when you launch your virtual machine:

Figure 1.11: SAS University Edition Start Menu

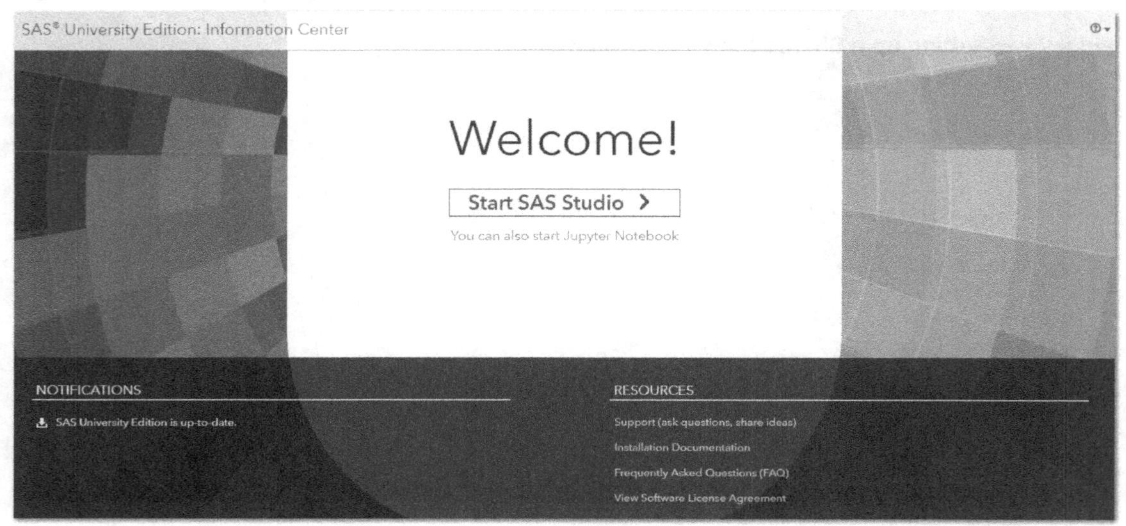

Be sure to click on the **Updates** button if updates are available (you may want to be sure you have a fully charged battery or have your computer plugged in). There are also links to installation documentation and other helpful resources.

You are now ready to start using SAS. When you close your SAS session, you can shut down (power off) or suspend your virtual computer. In most cases, you will want to suspend the machine (it takes longer to restart it if you power it down). When you are ready to start a new SAS session, you either power on the machine (if you turned it off) or resume (if you suspended the session).

The remaining chapters in the first section of this book show you how to operate in the point-and-click environment of SAS Studio—the second part of the book teaches you how to write SAS programs. Yes, there is a learning curve in working your way around SAS Studio and, especially, in learning to write SAS programs. A positive attitude is key. If you run into trouble, there are many sources to help you. You can Google almost any question concerning SAS Studio or SAS programming, and find a link to useful information.

Conclusion

Although it does take some work (and courage) to work in a virtual environment, especially if you have never done it before, the result is well worth the effort. There are many online resources that you can Google, and don't forget to find a friend (or son or daughter) who is knowledgeable about such things. You might consider taking them to lunch or, better yet, a nice dinner.

Chapter 2: The SAS Studio Interface

Introduction

To begin using the SAS University Edition, you need to enter an IP address in your browser. The form of this address will depend on which virtualization software you are using. For example, using VMware, you will be instructed to enter an address such as `http://192.168.117.129` (VMware will tell you which address to use). With VirtualBox, the address will be in the form: `http://localhost:10080` (VirtualBox will provide this address.) You should bookmark this location for future use.

When you first open SAS University Edition, you will see the following screen:

Figure 2.1: Starting SAS University Edition

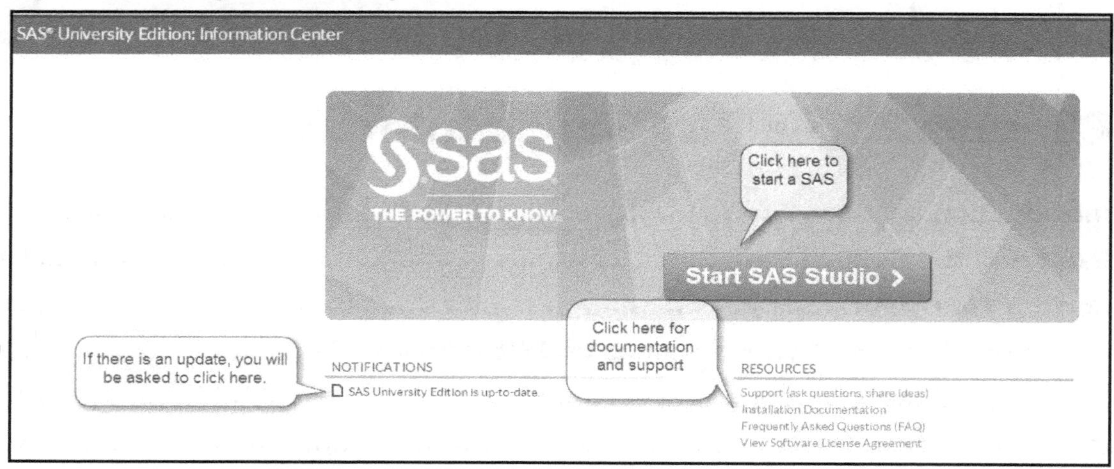

At this point, you can click on **Start SAS Studio**. There are also links to resources such as videos to give you further instructions. If the **Notifications** box shows that updates are available, you can request the updates or wait for another time to update your software.

The opening screen in SAS Studio looks like this:

Figure 2.2: Opening Screen for SAS Studio

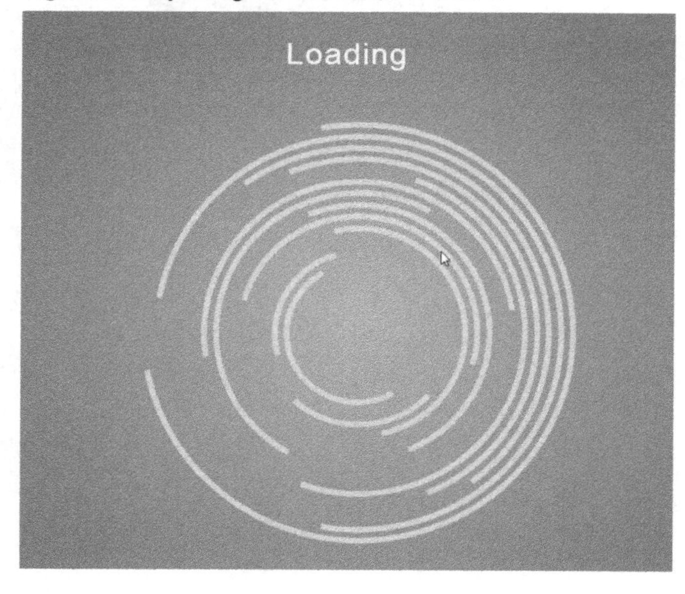

Don't stare at this —SAS is trying to hypnotize you!

After a few seconds, you arrive at the home screen (yours may vary because this author has already created files and folders):

Figure 2.3: Home Screen for SAS Studio

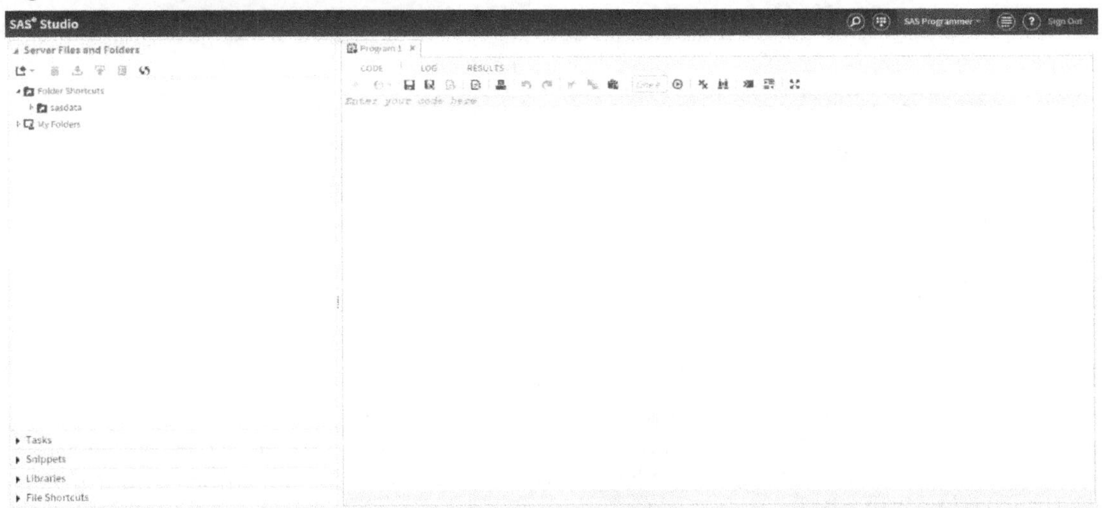

On the left, you see the navigation pane—on the right, the work area. Here is a blow up of the navigation pane:

Figure 2.4: Blow Up of the Navigation Pane

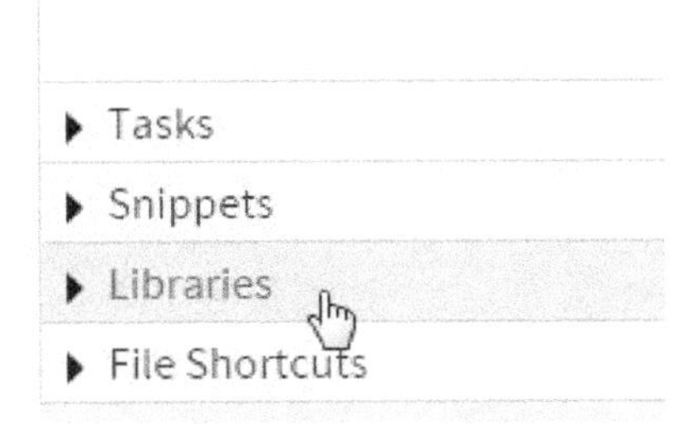

Exploring the Built-In Data Sets

SAS libraries are where SAS stores SAS data sets. SAS Studio ships with a number of data sets that you can play with. In the next chapter, you will see how to create your own SAS data sets from several different data sources and store them in a library of your own.

When you click on the **My Libraries** tab, you will see something like this:

Figure 2.5: Expanding the My Libraries Tab

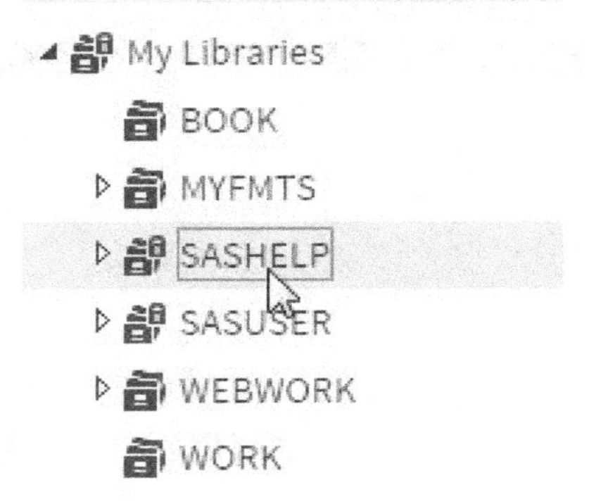

Some of the libraries you see here were created by the author. Others, such as Sasuser and Sashelp, will always show up. The Sashelp library is where SAS has stored all of the demonstration data sets. Click on the small triangle to the left of Sashelp to see a list of the SAS data sets stored there. It should look like this:

Figure 2.6: Expanding the Sashelp Library

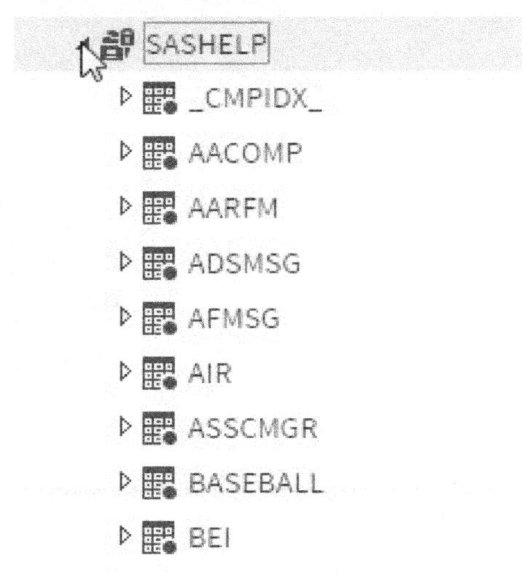

You can click on the small triangle to the left of any of these data sets to see a list of variables. As an alternative, double-click on a data set of interest to display a list of variables and a partial listing of the data set. For this example, let's double-click on the **BASEBALL** data set. Here's what happens:

Figure 2.7: Opening the BASEBALL Data Set

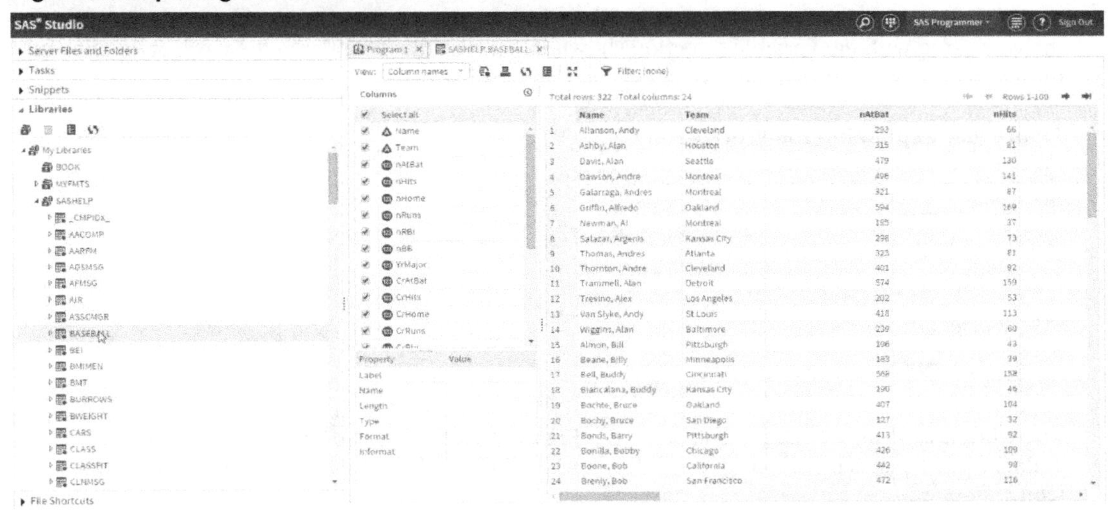

The middle pane shows a list of variables while the right pane shows a partial listing of the data. For a better view of the list of variables, the next figure shows an expanded view:

Figure 2.8: Expanded View of the Variables Pane

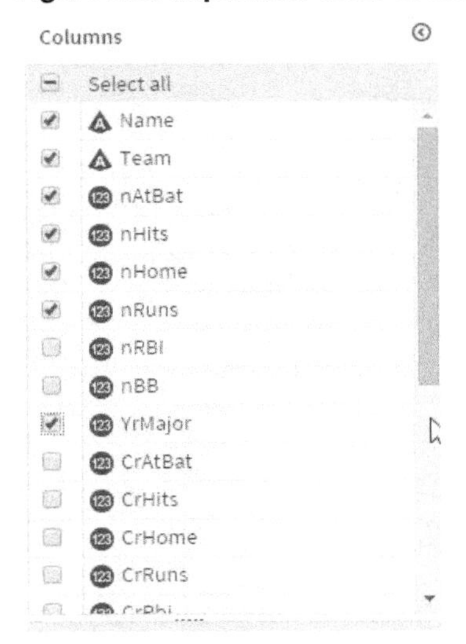

The default action is to select all the variables. You can click on any variable to unselect it or, alternatively, uncheck **Select All** and then, holding down the Ctrl key, select the variables you want to display. If the variables you wish to select are in sequence, you can also click on the first variable of interest, hold the Shift key down, and then click on the last variable in the list to select all the variables from the first to the last.

As you select or unselect variables, the work area changes to reflect these selections. Let's look at the work area with the variable selection shown in Figure 2.8. To enlarge the work area, click on the **Expand** icon to (shown below) to expand this area.

Figure 2.9: Expanding the Work Area

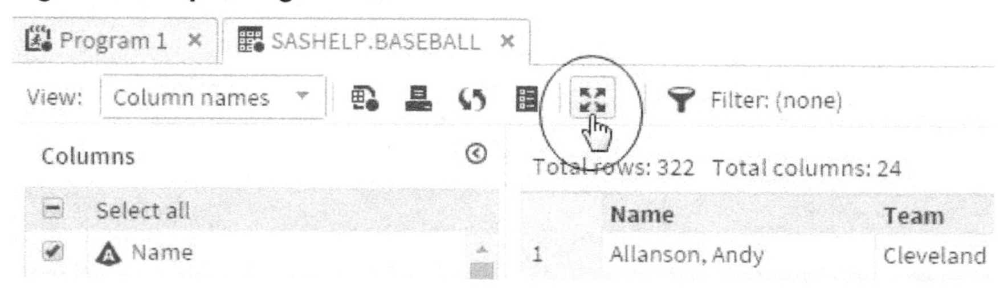

The work area is now enlarged and shown in Figure 2.10 below, (**Note:** You can expand or collapse other panes by clicking on the **Expand** icon or, if already expanded, on the **Collapse** icon.)

Figure 2.10: Work Area with Selected Variables

Total rows: 322 Total columns: 24 Rows 1-100

	Name	Team	nAtBat	nHits	nHome	nRuns
1	Aldrete, Mike	San Francisco	216	54	2	27
2	Allanson, Andy	Cleveland	293	66	1	30
3	Almon, Bill	Pittsburgh	196	43	7	29
4	Anderson, Dave	Los Angeles	216	53	1	31
5	Armas, Tony	Boston	425	112	11	40
6	Ashby, Alan	Houston	315	81	7	24
7	Backman, Wally	New York	387	124	1	67
8	Baines, Harold	Chicago	570	169	21	72
9	Baker, Dusty	Oakland	242	58	4	25
10	Balboni, Steve	Kansas City	512	117	29	54
11	Bando, Chris	Cleveland	254	68	2	28
12	Barfield, Jesse	Toronto	589	170	40	107
13	Barrett, Marty	Boston	625	179	4	94
14	Bass, Kevin	Houston	591	184	20	83
15	Baylor, Don	Boston	585	139	31	93
16	Beane, Billy	Minneapolis	183	39	3	20
17	Bell, Buddy	Cincinnati	568	158	20	89
18	Bell, George	Toronto	641	198	31	101
19	Belliard, Rafael	Pittsburgh	309	72	0	33
20	Beniquez, Juan	Baltimore	343	103	6	48
21	Bernazard, Tony	Cleveland	562	169	17	88
22	Biancalana, Buddy	Kansas City	190	46	2	24
23	Bilardello, Dann	Montreal	191	37	4	12
24	Bochte, Bruce	Oakland	407	104	6	57

You can use the scroll bars to scroll left or right, up or down. At the top of the work area, you see the number of rows (observations) and columns (variables) in the data set. Here is an enlarged view:

Figure 2.11: Number of Rows and Columns

You see that there are 322 rows and 24 columns. These numbers will change as you change your variable selections or create filters.

Sorting Your Data

If you place the cursor on a column heading, a small hand icon appears (see Figure 2.12).

Figure 2.12: Placing the Cursor on a Column Head

nAtBat
279
236
166
161
439
237
209
298
309
239
600
514

By left-clicking here (the column heading nAtBat stands for the number of times at bat), the values are sorted from low to high (an ascending sort). See Figure 2.13 below:

Figure 2.13: Ascending Sort

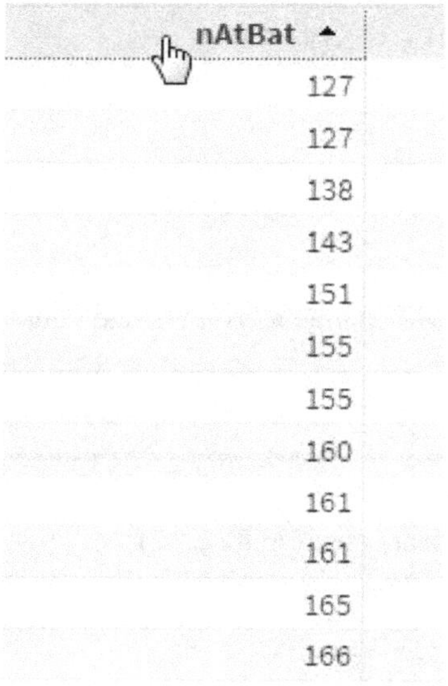

nAtBat ▲
127
127
138
143
151
155
155
160
161
161
165
166

You can left-click on this column again to request a descending sort (see Figure 2.14):

Figure 2.14: Descending Sort

nAtBat
687
680
677
663
642
641
637
633
631
629
627
627

Switching Between Column Names and Column Labels

You can switch between variable names and variable labels by clicking on the **View** tab.

Figure 2.15: Selecting Variable Names or Labels

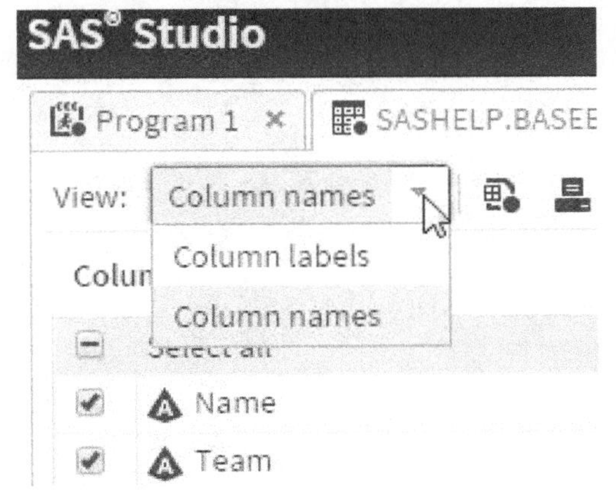

If you choose column labels, the column names are replaced by labels. (This only works, of course, if the person creating the data set created labels for the variables.) Figure 2.16 (below) shows the effect of switching to column labels for the Baseball data set (that did contain labels):

Figure 2.16: Column Headings Changed to Labels

Total rows: 322 Total columns: 24

	Player's Name	Team at the End of 1986	Times at Bat in 1986 ▾	Hits in 1986
1	Fernandez, Tony	Toronto	687	213
2	Puckett, Kirby	Minneapolis	680	223
3	Mattingly, Don	New York	677	238
4	Carter, Joe	Cleveland	663	200
5	Gwynn, Tony	San Diego	642	211
6	Bell, George	Toronto	641	198
7	[illegible]	[illegible]	[illegible]	[illegible]

Resizing Tables

By placing the cursor on the dividing line between columns, the hand icon changes to double vertical lines, allowing you to then drag the border of the column left or right. In Figure 2.17 below, the nAtBat column was resized (made smaller):

Figure 2.17: Resizing a Column

Team	nAtBat
San Diego	127
Atlanta	127
Philadelphia	138
San Francisco	143
California	151
Chicago	155
Texas	155
Baltimore	160
New York	161
Oakland	161
Minneapolis	165

Creating Filters

Right-clicking on a column brings up the following menu:

Figure 2.18: Adding a Filter

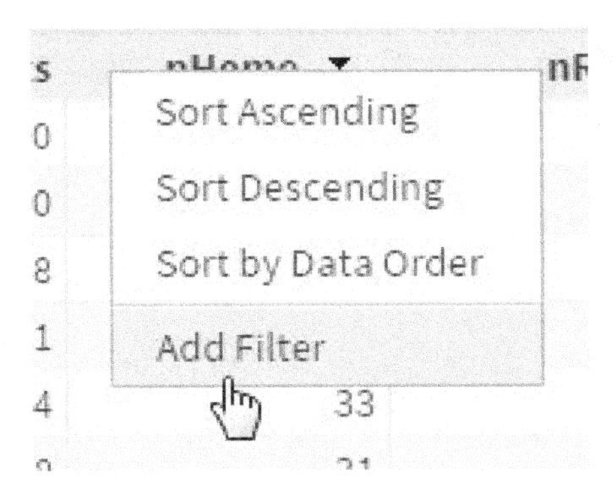

Here you see that you can perform ascending or descending sorts (the same as left-clicking on a column head), you can sort in data order, and you can add a filter. *Filters* allow you to subset rows of the table by specifying criteria, such as displaying only rows where the number of home runs was 30 or more. The following screen shows how to create a filter. Click on **Add Filter** to display the screen shown next:

Figure 2.19: Adding a Filter (Continued)

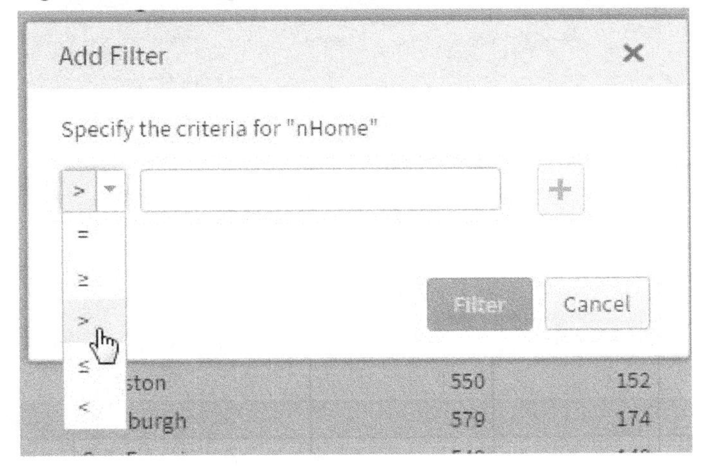

In the drop-down menu, select a logical operator. In this example, you are choosing greater than:

Figure 2.20: Adding a Filter (Continued)

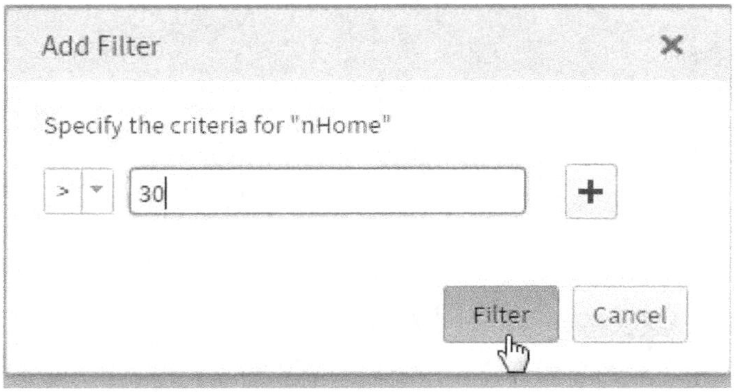

Add Filter ×

Specify the criteria for "nHome"

> 30 +

Filter Cancel

Next, enter a criterion for the variable you have selected. If you choose a categorical variable like Team, then the filter dialog gives you the list of values that you can choose from. If you wish to add more conditions, click on the + (plus) sign. If you are finished, click on **Filter** to complete your request.

Figure 2.21: Listing of Filtered Data

Total rows: 322 Total columns: 24 Filtered rows: 11

	Name	Team	nAtBat	nHits	nHome
1	Barfield, Jesse	Toronto	589	170	40
2	Schmidt, Mike	Philadelphia	552	160	37
3	Kingman, Dave	Oakland	561	118	35
4	Gaetti, Gary	Minneapolis	596	171	34
5	Canseco, Jose	Oakland	600	144	33
6	Mattingly, Don	New York	677	238	31
7	Davis, Glenn	Houston	574	152	31
8	Baylor, Don	Boston	585	139	31
9	Bell, George	Toronto	641	198	31
10	Puckett, Kirby	Minneapolis	680	223	31
11	Parker, Dave	Cincinnati	637	174	31

Here you see all rows in the Baseball data set where the player hit more than 30 home runs.

If you wish to remove the filter, click on the **X** next to the filter, as shown in Figure 2.22:

Figure 2.22: Removing a Filter

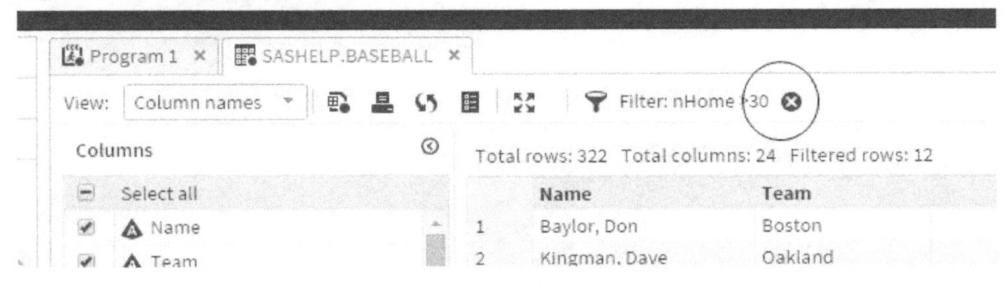

Conclusion

In this chapter, you saw how to open one of the built-in Sashelp data sets and the various tasks that you can perform using simple point-and-click operations supplied by SAS Studio. The next step is to create SAS data sets using your own data and then perform operations such as summarizing your data; creating plots and charts; and creating reports in HTML, PDF, or RTF format. You will learn how to accomplish all of these operations in the chapters to follow.

Chapter 3: Importing Your Own Data

Introduction

In the last chapter, you saw how to use some of the features of SAS Studio to manipulate data from the built-in SAS data sets in Sashelp. In this chapter, you will see how easy it is to import your own data from Excel workbooks, CSV files, and many other file formats such as Access and SPSS to create SAS data sets.

Exploring the Utilities Tab

If you open up your **Tasks** tab in the navigation pane, you will see the following:

Figure 3.1: The Tasks Tab in the Navigation Pane

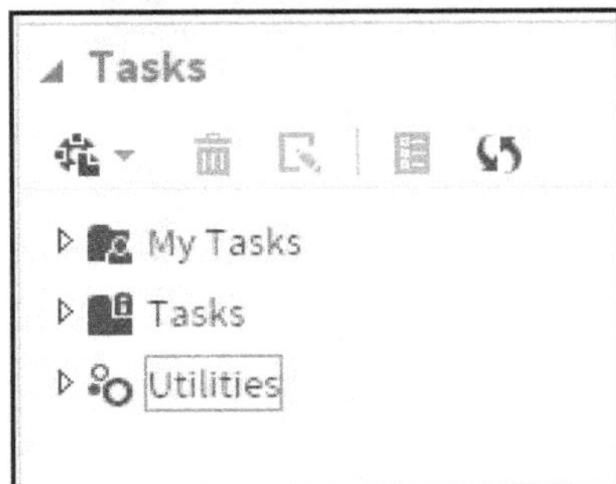

Now, expand the **Utilities** tab. It looks like this:

Figure 3.2: Expanding the Utilities Tab

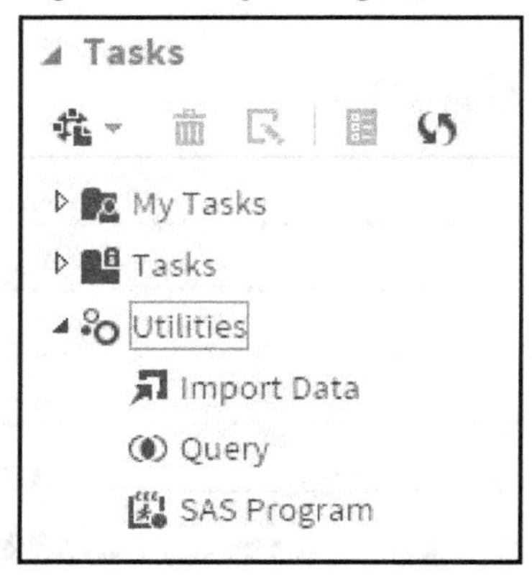

The **Import Data** task is used to import data in a variety of formats and create SAS data sets. A complete list of supported file types is shown in Figure 3.3 below:

Figure 3.3: List of Supported Files

DEFAULT (Based on file extension)
ACCESS (Microsoft Access using LIBNAME statement)
CSV (Comma delimited file)
DBF (dBASE 5.0, IV, III+ and III)
DBFMEMO (dBASE 5.0, IV, III+ and III with memos)
DLM (Delimited file)
DTA (Stata file)
EXCEL (Microsoft Excel using LIBNAME statement)
JMP (JMP file)
PARADOX (Paradox DB)
SPSS (SPSS file)
WK1 (Lotus 1-2-3 Release 2)
WK3 (Lotus 1-2-3 Release 3)
WK4 (Lotus 1-2-3 Release 4 or 5)
XLS (Microsoft Excel 5.0, 95, 97, 00-03)
XLSX (Microsoft Excel 2007 or later workbook)

As you can see, this import data utility can import data from many of the most common data formats. Because Excel workbooks and CSV files are so popular, let's use them to demonstrate how SAS converts various file formats into SAS data sets.

The Excel selection can accept both the older XLS workbooks as well as the current form with XLSX extensions. As you will see, the import data task can use file extensions to automatically determine the file type. If you have a non-standard file extension or if you prefer to manually select a file format, you can use the drop-down list in Figure 3.3 to instruct SAS on how to convert your file.

Importing Data from an Excel Workbook

Let's use the workbook Grades.xlsx (located in the folder `SASUniversityEdition/myfolders`) for this demonstration. If you open this workbook in Excel, it looks like this:

Figure 3.4: Excel Workbook Grades.xlsx

	A	B	C	D	E	F	G	H
1	Name	ID	Quiz1	Quiz2	Midterm	Quiz3	Quiz4	Final
2	Jones	12345	88	80	76	88	90	82
3	Hildebran	22222	95	92	91	94	90	96
4	O'Brien	33333	76	78	79	81	83	80
5								

The first row of the worksheet contains variable names (also known as *column names*). The remaining rows contain data on three students (yup, it was a very small class). The worksheet name was not changed, so it has the default name Sheet1.

The first step to import this data into a SAS data set is to double-click on the **Import Data** task.

Figure 3.5: Double-Clicking on the Import Data Task

You have two ways to select which file you want to import. One is to click on the **Select File** button on the right side of the screen—the other method is to click on the **Server Files and Folders** tab in the navigation pane (on the left), find the file, and drag it to the drag and drop area.

Clicking on **Select File** brings up a window where you can select a file to import. Here it is:

Figure 3.6: Clicking on the Select File Button

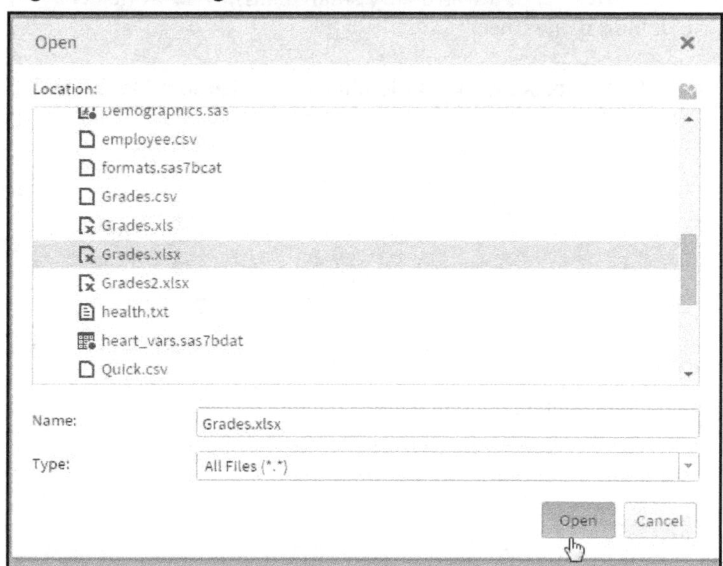

Select the file you want to import and click on **Open**. This brings up the Import window:

Figure 3.7: The Import Window

The top part of the window shows information on the file you want to import. You can enter a worksheet name (if there are multiple worksheets) and change the default SAS data set name by clicking on the **Change** button. You can use the mouse to enlarge the top half of the Import window or use the scroll bar on the right to reveal the entire window. The figure below shows the expanded view of the Import window:

Figure 3.8: Expanded View of the Import Window

Because you only have one worksheet, you do not have to enter a worksheet name. You probably want to change the name of the output (SAS) data set. Clicking on the **Change** button brings up a list of SAS libraries (below):

Figure 3.9: Changing the Name of the SAS Data Set

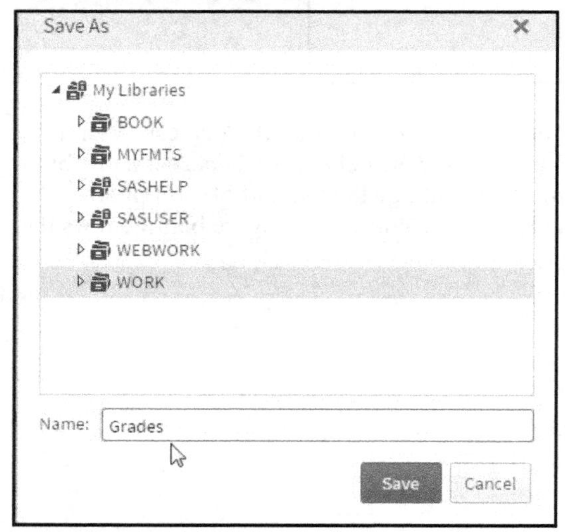

The Work library is used to create a temporary SAS data set (that disappears when you close your SAS Session). For now, let's select the Work library and name the data set Grades.

The **OPTIONS** pull-down menu allows you to select a file type. However, if your file has the appropriate extension (for example, XLSX, XLS, or CSV), you can leave the default actions (based on the file extension) to decide how to import the data.

Because the first row of the spreadsheet contains variable names, leave the check on the **Generate SAS variable names** option. This tells the import utility to use the first row of the worksheet to generate variable names.

When all is ready, click on the **Run** icon (Figure 3.10 below).

Figure 3.10: Click on the Run Icon

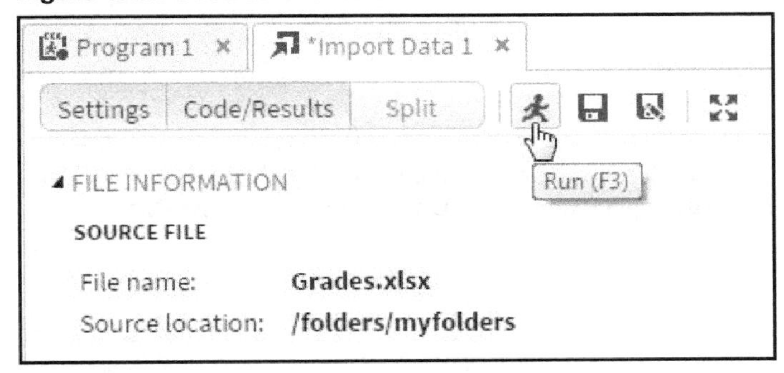

You are done! Here's what the screen looks like:

Figure 3.11: Variable List for the Work.Grades SAS Data Set

Alphabetic List of Variables and Attributes						
#	Variable	Type	Len	Format	Informat	Label
8	Final	Num	8	BEST.		Final
2	ID	Num	8	BEST.		ID
5	Midterm	Num	8	BEST.		Midterm
1	Name	Char	10	$10.	$10.	Name
3	Quiz1	Num	8	BEST.		Quiz1
4	Quiz2	Num	8	BEST.		Quiz2
6	Quiz3	Num	8	BEST.		Quiz3
7	Quiz4	Num	8	BEST.		Quiz4

Here you see a list of the variable names, whether they are stored as numeric or character, along with some other information that we don't need at this time. Notice that the import utility correctly read Name as character and the other variables as numeric.

Listing the SAS Data Set

A quick way to see a listing of the Grades data set is to select the **My Libraries** tab in the navigation pane, open up the Work library, and double-click on **Grades**. It looks like this:

Figure 3.12: Data Set Grades in the Work Library

Total rows: 3 Total columns: 8 Rows 1-3

	Name	ID	Quiz1	Quiz2
1	Jones	12345	88	80
2	Hildebrand	22222	95	92
3	O'Brien	33333	76	78

You can use your mouse to scroll to the right to see the rest of the table. To create a nicer looking report, click on the **Task** tab of the navigation pane and select **Data** followed by **List Data**, like this:

Figure 3.13: The List Data Task

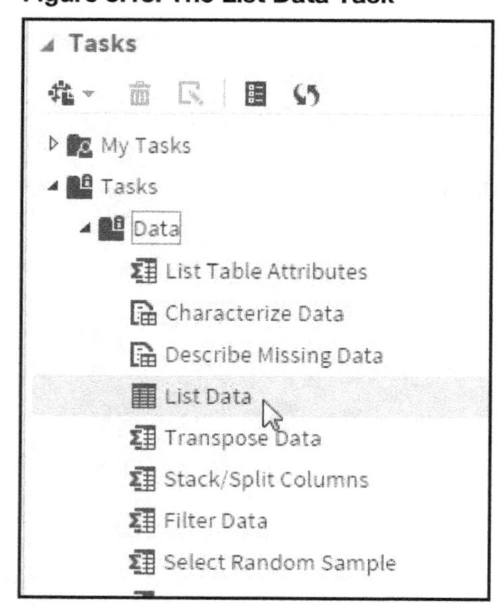

Double-click on **List Data**, and select the **Grades** data set in the **Work** library, and then click on the **Run** icon. You will be presented with a nice-looking listing of the Grades data set (see Figure 3.14 below):

Figure 3.14: Listing of the Grades Data Set

List Data for WORK.GRADES

Obs	Name	ID	Quiz1	Quiz2	Midterm	Quiz3	Quiz4	Final
1	Jones	12345	88	80	76	88	90	82
2	Hildebrand	22222	95	92	91	94	90	96
3	O'Brien	33333	76	78	79	81	83	80

Importing an Excel Workbook with Invalid SAS Variable Names

What if your Excel worksheet has column headings that are not valid SAS variable names?

> Valid SAS data set names are up to 32 characters long. The first character must be a letter or underscore—the remaining characters can be letters or digits. You are free to use upper- or lowercase letters.

As an example, take a look at the worksheet Grades2 shown in Figure 3.15:

Figure 3.15: Listing of Excel Workbook Grades2

	A	B	C	D	E	F	G	H
1	Stuent Name	ID	Quiz 1	Quiz 2	Mid Term	Quiz 3	Quiz 4	2015Final
2	Jones	12345	88	80	76	88	90	82
3	Hildebrand	22222	95	92	91	94	90	96
4	O'Brien	33333	76	78	79	81	83	80
5								

Most of the column headings in this spreadsheet are not valid SAS variable names. Six of them contain a blank and the last column (2015Final) starts with a digit. What happens when you import this worksheet? Because you now know how to use the **Import Data** task, it is not necessary to describe the import task again. All you really need to see is the final list of variables in the data set. Here they are:

Figure 3.16: Variable Names in the Grades2 SAS Data Set

Alphabetic List of Variables and Attributes						
#	Variable	Type	Len	Format	Informat	Label
2	ID	Num	8	BEST.		ID
5	Mid_Term	Num	8	BEST.		Mid Term
3	Quiz_1	Num	8	BEST.		Quiz 1
4	Quiz_2	Num	8	BEST.		Quiz 2
6	Quiz_3	Num	8	BEST.		Quiz 3
7	Quiz_4	Num	8	BEST.		Quiz 4
1	Stuent_Name	Char	10	$10.	$10.	Stuent Name
8	_2015Final	Num	8	BEST.		2015Final

As you can see, SAS replaced all the blanks with underscores and added an underscore as the first character in the 2015Final name to create valid SAS variable names.

Importing an Excel Workbook That Does Not Have Variable Names

What if the first row of your worksheet does not contain variable names (column headings)? You have two choices: First, you can edit the worksheet and insert a row with variable names. The other option is to uncheck **Generate SAS variable names** in the **OPTIONS** section in the Import window (see Figure 3.17), and let SAS create variable names for you.

Figure 3.17: Uncheck the Generate SAS Variable Names Option

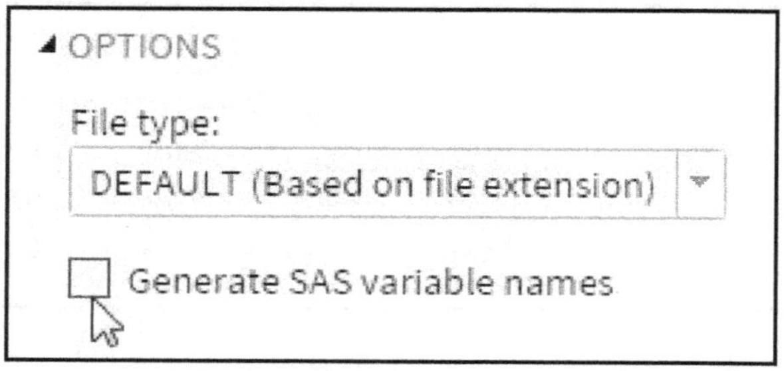

Here is the result:

Figure 3.18: SAS Generated Variable Names

Alphabetic List of Variables and Attributes						
#	Variable	Type	Len	Format	Informat	Label
1	A	Num	8	BEST.		A
2	B	Char	1	$1.	$1.	B
3	C	Num	8	MMDDYY10.		C
4	D	Num	8	BEST.		D
5	E	Num	8	BEST.		E
6	F	Char	1	$1.	$1.	F

SAS used the column identifiers (A through F) as variable names. You can leave these variable names as is or change them using DATA step programming. Another option is to use PROC DATASETS, a SAS procedure that allows you to alter various attributes of a SAS data set without having to create a new copy of the data set.

When you import a CSV file without variable names, you will see SAS generated variable names VAR1, VAR2, etc.

Importing Data from a CSV File

CSV (comma-separated values) files are a popular format for external data files. As the name implies, CSV files use commas as data delimiters. Many web sites allow you to download data as CSV files. As with Excel workbooks, your CSV file may or may not contain variable names at the beginning of the file. If the file does contain variable names, be sure the **Generate SAS variable names** options box is checked; if not, uncheck this option.

As an example, take a look at the CSV file called Grades.csv in Figure 19 below:

Figure 3.19: CSV File Grades.csv

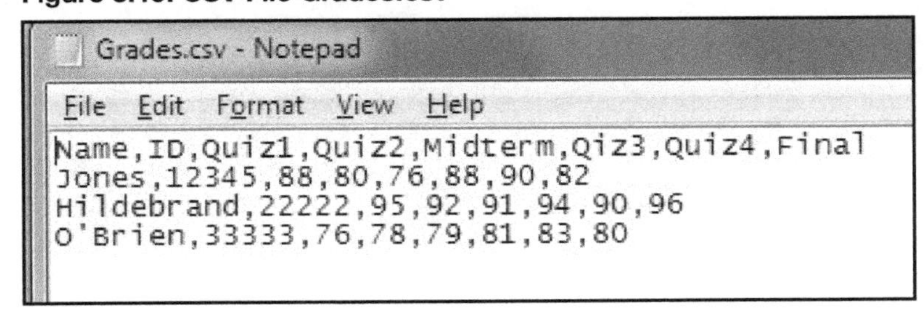

Grades.csv - Notepad
File Edit Format View Help
Name,ID,Quiz1,Quiz2,Midterm,Qiz3,Quiz4,Final
Jones,12345,88,80,76,88,90,82
Hildebrand,22222,95,92,91,94,90,96
O'Brien,33333,76,78,79,81,83,80

This CSV file contains the same data as the Excel workbook Grades.xlsx. Notice that variable names are included in the file. You can import this file and create a SAS data set, using the same steps you used to import the Excel workbook. The import facility will automatically use the correct code to import this data file because of the CSV file extension. The resulting SAS data set is identical to the one shown in Figure 3.14.

Conclusion

You have seen how easy it is to import data in a variety of formats and create SAS data sets. Even experienced programmers (at least the ones this author knows) would prefer to use the import data utility to convert external data to SAS data sets rather than writing their own code.

When you learn how to access data in folders other than the `Myfolders` folder (the default file location for SAS Studio), you can import data from anywhere on your hard drive.

Chapter 4: Creating Reports

Introduction

In the last chapter, you saw how to use the **Import Data** facility on the **Utilities** tab to import data. In this chapter, you will see how to use several of the most useful tasks as well as the Query tool on the **Utilities** tab.

> In this book, as well as in SAS Studio, you will see the terms SAS data set and table used interchangeably as well as these other equivalent terms: variables are also called columns and observations are called rows.

Using the List Data Task to Create a Simple Listing

SAS Studio has dozens of built-in tasks. If you expand the **Tasks** tab, you will see the following:

Figure 4.1: Data Tasks

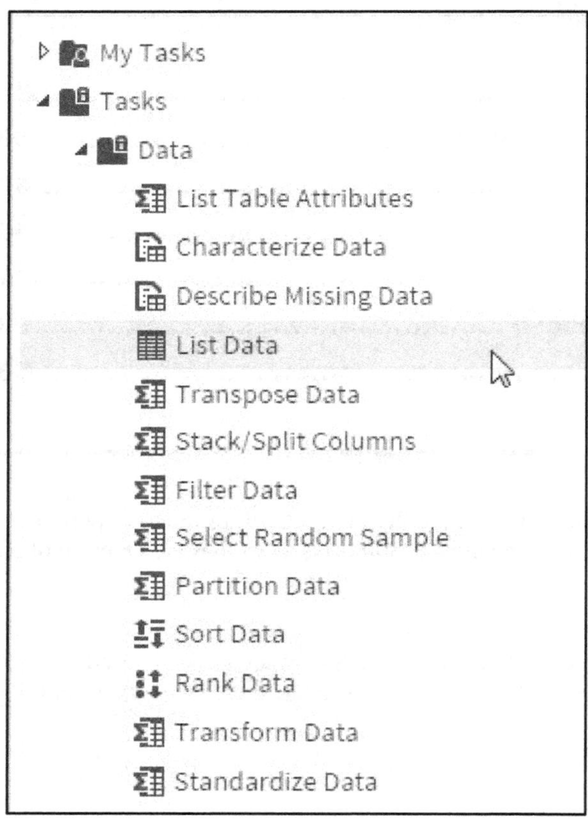

You can use the **List Data** task to create a listing of your data set. To demonstrate this, we are going to use the built-in Sashelp data set called Fish. This data set contains information on several species of fish, including weight, length, and width. To create a listing of this data set, expand the list of **Data** tasks and double-click on **List Data**. This brings up the screen shown in Figure 4.2:

Figure 4.2: The List Data Task Settings Screen

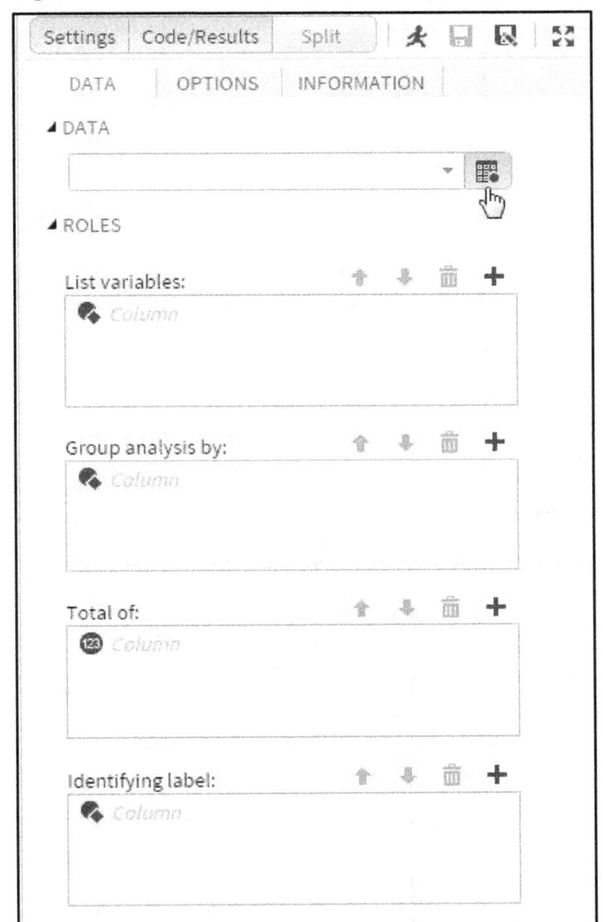

You can click on the icon at the top-right part of this screen to select the library and data set you wish to list. Because you want a listing of Sashelp.Fish, select this data set.

Figure 4.3: Selecting the Fish Data Set in the Sashelp Library

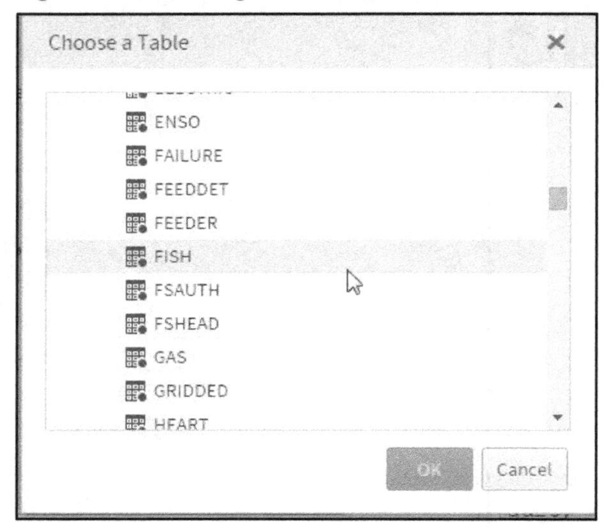

The next step is to click on the plus sign (+) to select which variables you want to include in your listing (see Figure 4.4):

Figure 4.4: Adding Columns (Variables)

When you see the list of variables, you can select them in the usual way (see the instructions below):

> To select variables from a list, use one of these two methods: 1) Hold the Ctrl key down and select the variables you want; or 2) click on one variable, hold the Shift key down, and click on another variable—all the variables from the first to the last will be selected.

In this example, you are selecting **Species**, **Weight**, **Height**, and **Width**.

Figure 4.5: Selecting Variables to List

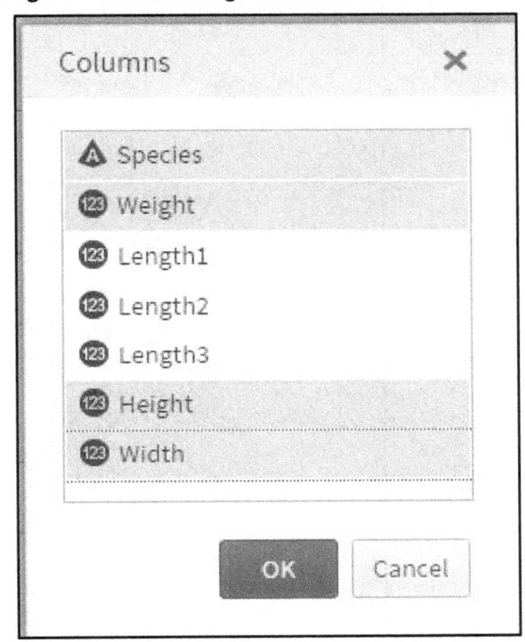

Click **OK** when you are finished. You can create the listing now or click on the **OPTIONS** tab to customize the listing.

Figure 4.6: Using the OPTIONS Tab to Customize the Listing

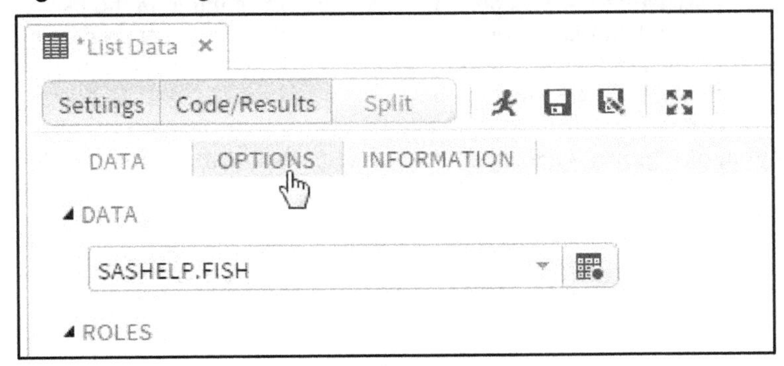

Here is the list of options available for the **List Data** task:

Figure 4.7: Options for the List Data Task

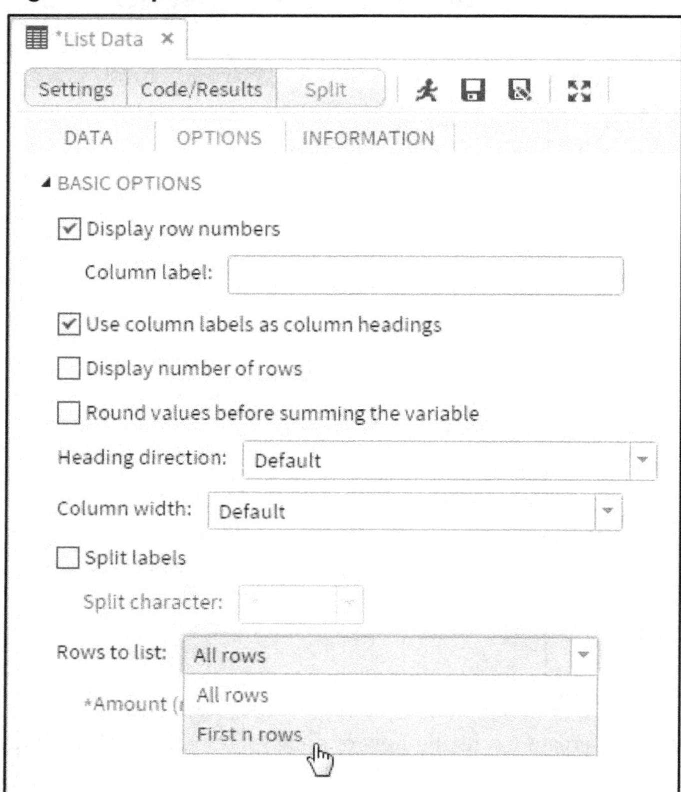

You can check or uncheck the **Display row numbers** box to include the Obs column in the listing or omit it. You have a choice of using column labels or column names in the listing. If you are a programmer, you will probably want to see column names as headings—if you are creating the listing for a report, you will probably want to see column labels.

The option displayed at the bottom of Figure 4.7 gives you the choice of listing all the rows of the table or the first *n* rows. In this example, you want to see the first seven rows of the Fish data set. This is shown in the next figure:

Figure 4.8: Requesting the First Seven Rows to Be Displayed

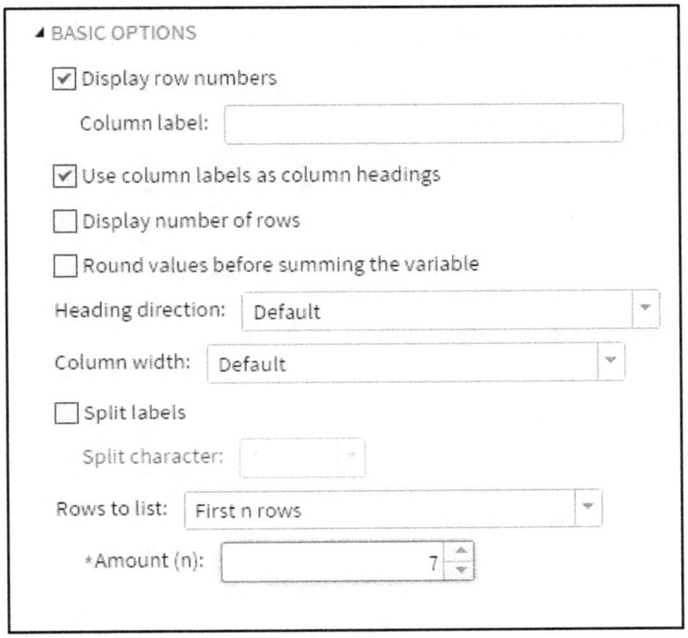

BASIC OPTIONS

[x] Display row numbers

Column label:

[x] Use column labels as column headings

[ ] Display number of rows

[ ] Round values before summing the variable

Heading direction: Default

Column width: Default

[ ] Split labels

Split character:

Rows to list: First n rows

*Amount (n): 7

Clicking on the **Run** icon generates the following listing:

Figure 4.9: Listing of the First Seven Rows of the Fish Data Set

List Data for SASHELP.FISH

Obs	Species	Weight	Height	Width
1	Bream	242	11.5200	4.0200
2	Bream	290	12.4800	4.3056
3	Bream	340	12.3778	4.6961
4	Bream	363	12.7300	4.4555
5	Bream	430	12.4440	5.1340
6	Bream	450	13.6024	4.9274
7	Bream	500	14.1795	5.2785

Filtering Data

Another useful task is to filter the table—that is, you select rows that meet predefined criteria. To do this, double-click on **Filter Data** in the task list.

Figure 4.10: Selecting Filter Data from the Task List

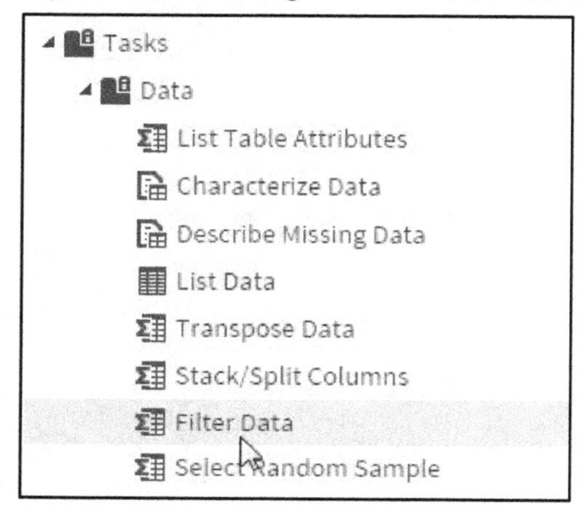

This brings up the following:

Figure 4.11: Selections for Filtering Data

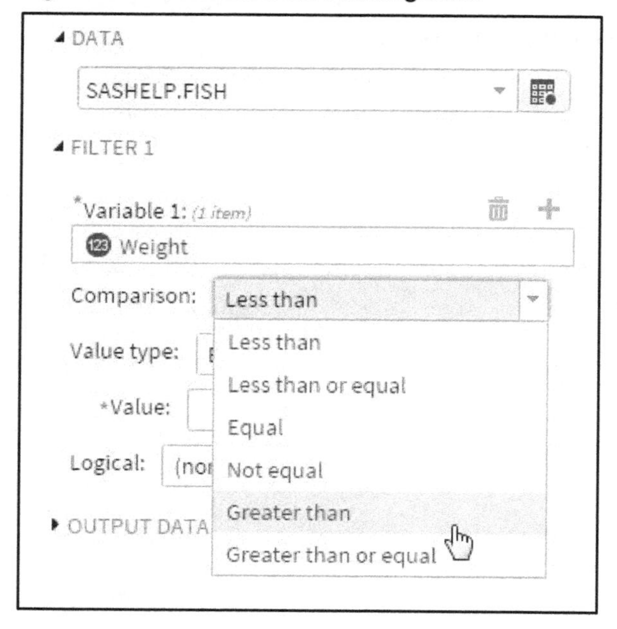

You select the data set as before. Next, you select a variable and a condition for your filter. In this example, you are selecting **Weight** as your variable and **Greater than** as your condition. You can now enter a value for the filter. In this example, you want to see rows in the table where the variable Weight is greater than 1,100.

Figure 4.12: Selecting Rows Where the Weight Is Greater Than 1,100

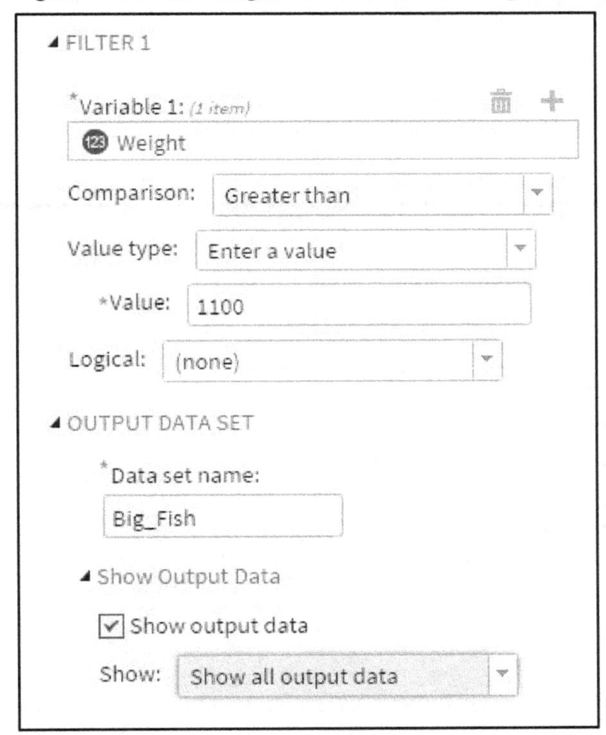

You can also expand the **OUTPUT DATA SET** option to override a default data set name. In most cases, you will want to supply your own data set name. Here you are naming the data set Big_Fish. You can also check or uncheck the **Show Output Data** box. Selecting it (as in this example) generates a listing of the filtered data set:

Figure 4.13: Listing of the Filtered Data Set (Big_Fish)

Filtered data set - WORK.Big_Fish

Obs	Species	Weight	Length1	Length2	Length3	Height	Width
1	Pike	1250	52	56.0	59.7	10.6863	6.9849
2	Pike	1600	56	60.0	64.0	9.6000	6.1440
3	Pike	1550	56	60.0	64.0	9.6000	6.1440
4	Pike	1650	59	63.4	68.0	10.8120	7.4800

Sorting Data

To sort data, select **Sort Data** from the list:

Figure 4.14: The Sort Data Task

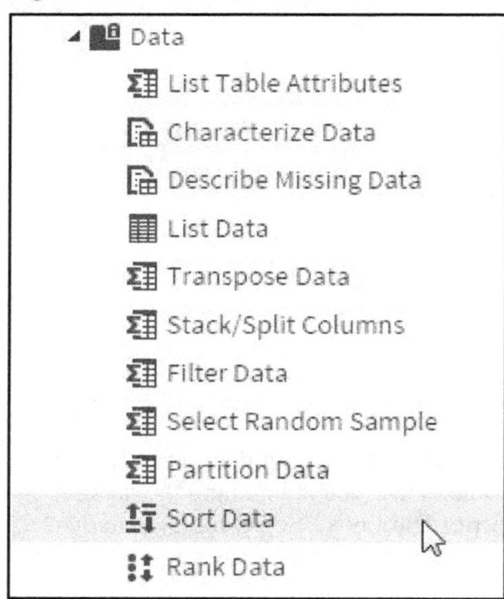

Just as in the previous tasks, you can now choose a data set and options.

Figure 4.15: Selecting a Data Set and Variables for the Sort

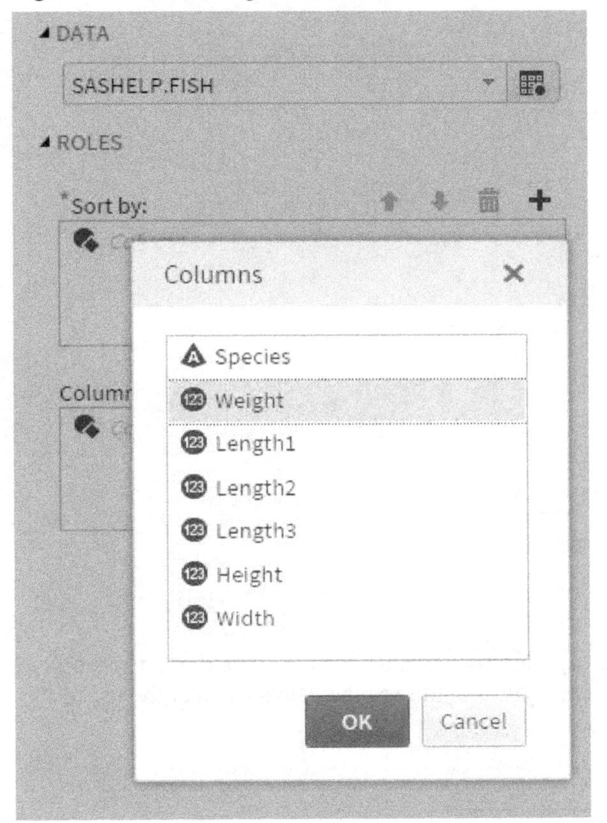

You are starting with the Sashelp.Fish data set and requesting a sort based on the variable Weight. You can also choose columns to drop:

Figure 4.16: Selecting Columns to Drop

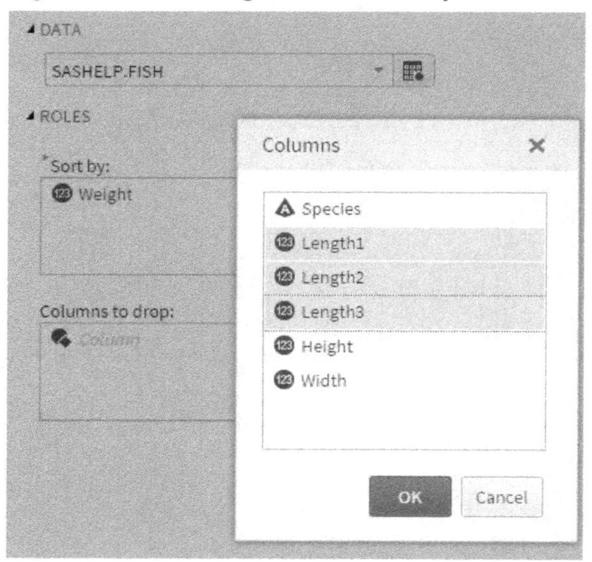

You are dropping the three Length variables.

Before you execute the sort, there are several options you should consider. The default sort order is ascending (from smallest to largest). In this case, you want to see the heavier fish at the top of the list, so you choose **Descending** as the sort order.

Figure 4.17: Selecting a Descending Sort

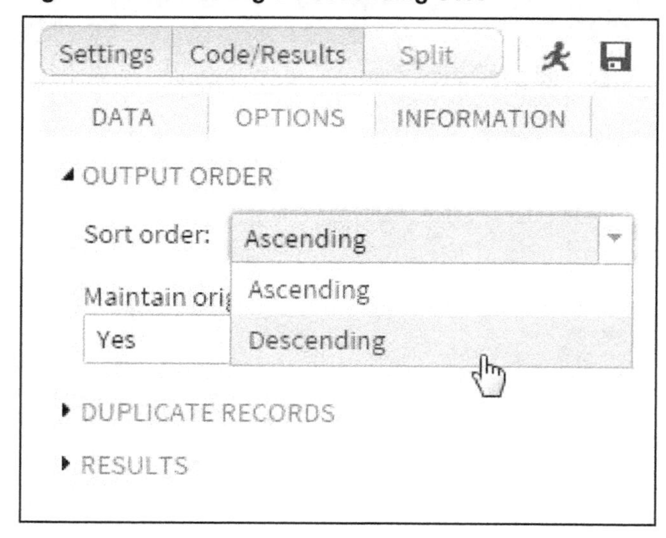

Expand the **RESULTS** option to either sort in place or create a new data set with the sorted data.

CAUTION: Sorting in place replaces the original data set with the sorted data. If you drop columns, they will no longer be in the sorted data set.

In this example, you want to create a new data set called Sorted_Fish:

Figure 4.18: Naming the Output Data Set

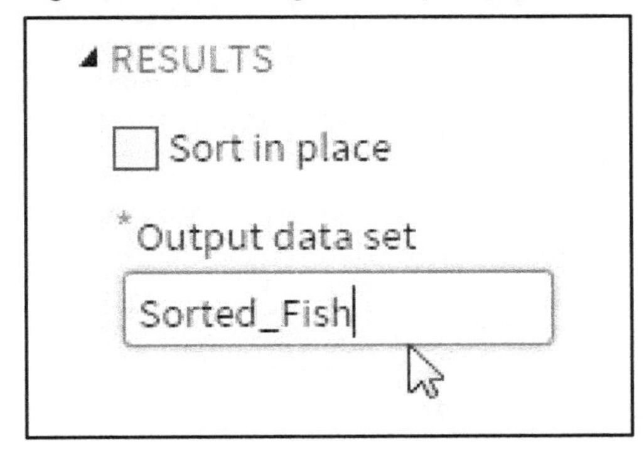

Click on the **Run** icon to see the following screen:

Figure 4.19: Result of Executing the Sort

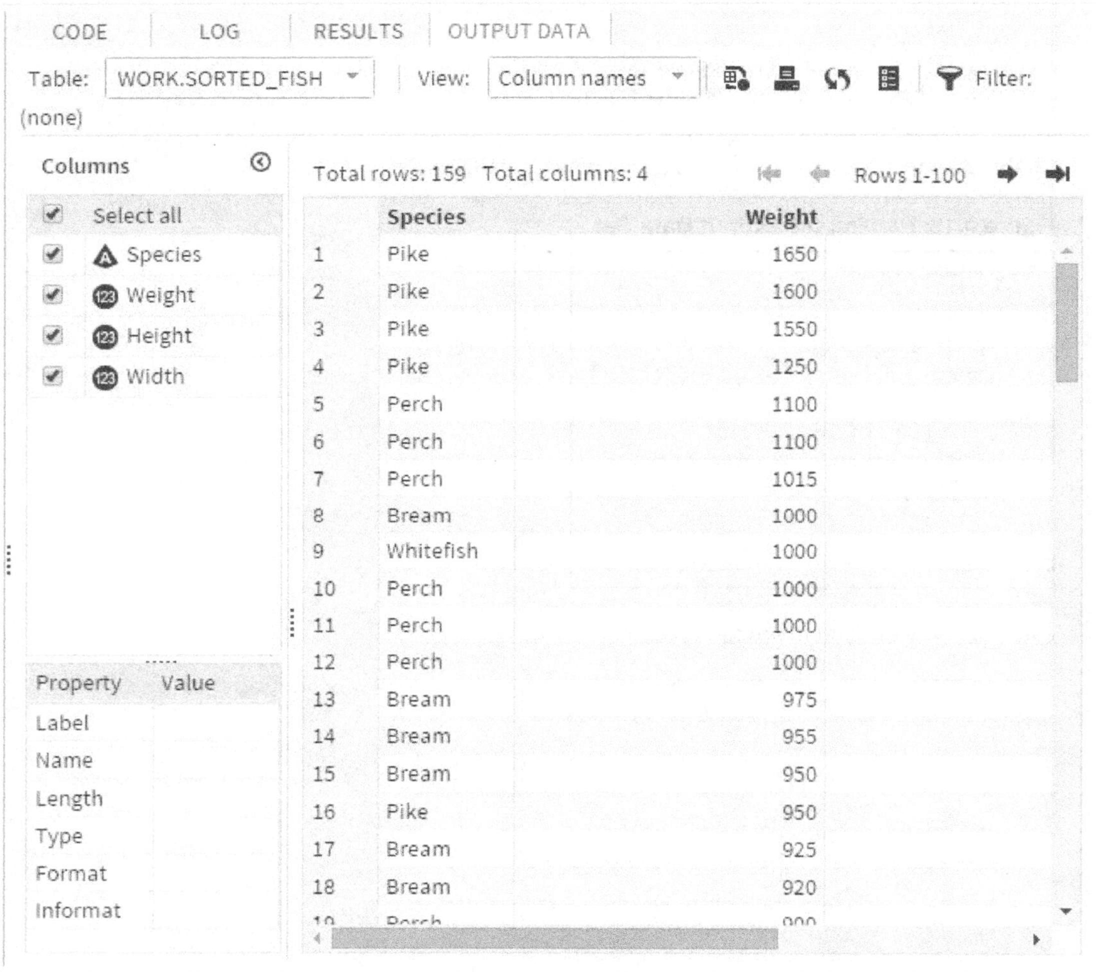

CODE LOG RESULTS OUTPUT DATA

Table: WORK.SORTED_FISH View: Column names Filter: (none)

Columns

Select all

Species

Weight

Height

Width

Property	Value
Label	
Name	
Length	
Type	
Format	
Informat	

Total rows: 159 Total columns: 4 Rows 1-100

	Species	Weight
1	Pike	1650
2	Pike	1600
3	Pike	1550
4	Pike	1250
5	Perch	1100
6	Perch	1100
7	Perch	1015
8	Bream	1000
9	Whitefish	1000
10	Perch	1000
11	Perch	1000
12	Perch	1000
13	Bream	975
14	Bream	955
15	Bream	950
16	Pike	950
17	Bream	925
18	Bream	920

To see a nicer listing of this data set, go back to the **List Data** selection and proceed as you did in the first section of this chapter. Once you have opened the **List Data** task, choose a table to display and any options that you want.

Figure 4.20: Choose a Table to List

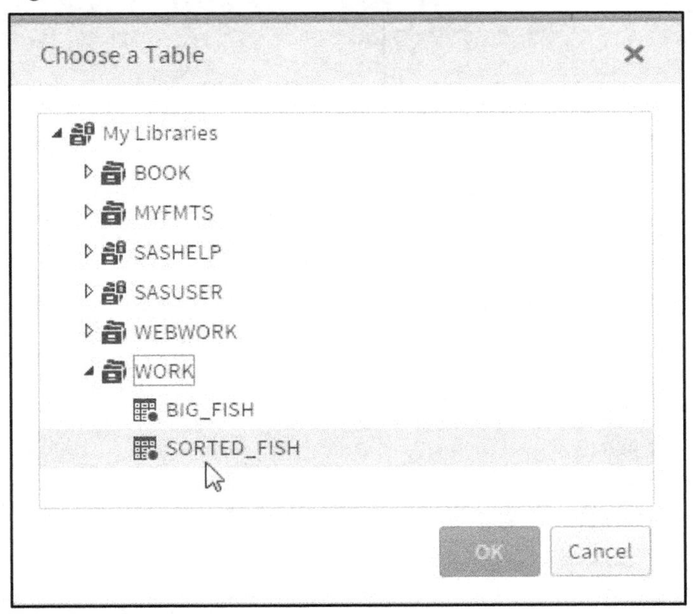

Next, choose which variables to include in the listing.

Figure 4.21: Choosing Variables to List

Finally, select any options that you want. In this example, you want to see the first eight rows of the table.

Figure 4.22: Option to List the First Eight Rows of the Table

Rows to list: First n rows

*Amount (n): 8

Here is the listing:

Figure 4.23: First Eight Rows of the Sorted_Fish Data Set

List Data for WORK.SORTED_FISH

Obs	Species	Weight	Height	Width
1	Pike	1650	10.8120	7.4800
2	Pike	1600	9.6000	6.1440
3	Pike	1550	9.6000	6.1440
4	Pike	1250	10.6863	6.9849
5	Perch	1100	12.8002	6.8684
6	Perch	1100	12.5125	7.4165
7	Perch	1015	12.3808	7.4624
8	Bream	1000	18.9570	6.6030

Outputting PDF and RTF Files

If you want either PDF or RTF (rich text format) output, first click on the SAS Studio **Options** icon.

Figure 4.24: Click on More Application Options

Click on **Preferences**, then **Results**, and then check or uncheck **PDF** and/or **RTF**. (Note: These preferences will remain in effect unless you change them later.)

If you have checked one of these file types, after you produce a listing, the two icons **Download results as a PDF file** or **Download results as an RTF file** will be displayed right above your listing (see Figure 4.25). Clicking on either of these icons will output the appropriate file type to a location of your choice.

Figure 4.25: Downloading a PDF File

CODE LOG RESULTS

List D Download results as a PDF file

Species	Weight	Height	Width
Bream	242	11.5200	4.0200
Bream	290	12.4800	4.3056
Bream	340	12.3778	4.6961
Bream	363	12.7300	4.4555
Bream	430	12.4440	5.1340

Here is a listing of the PDF file:

Figure 4.26: Listing of the PDF File

List Data for SASHELP.FISH

Species	Weight	Height	Width
Bream	242	11.5200	4.0200
Bream	290	12.4800	4.3056
Bream	340	12.3778	4.6961
Bream	363	12.7300	4.4555
Bream	430	12.4440	5.1340

Joining Tables (Using the Query Window)

The last topic in this chapter describes how to use the **Query** utility to join two tables. Two data sets, ID_Name and Grades, were created to explain how the joining process works. Here is a listing of these two data sets:

Figure 4.27: Listing of Data Sets ID_Name and Grades

List Data for WORK.ID_NAME

ID	Name
001	Ron
002	Jan
003	Peter
004	Paul
005	Mary

List Data for WORK.GRADES

ID	Grade1	Grade2	Grade3
005	78	80	82
002	100	90	95
001	99	95	98
006	65	67	69
004	85	86	84

There are several features of these two data sets that are important for you to notice. First, the ID variable in the ID_Name data set is in order—in the data set Grades, it is not. As you will see, this does not cause a problem—the Query tool automatically sort the data sets. Also, ID 003 is in data set ID_Name and not in data set Grades; ID 006 is in Grades but not in ID_Name. The goal is to join these two tables based on the ID column.

If you want to play along with this demonstration, you can run the program shown next to create these two tables (you may need to refer to the programming chapters to see how to do this):

Figure 4.28: Program to Create Data Sets ID_Name and Grades

```
data ID_Name;
   informat ID $3. Name $12.;
   input ID Name;
datalines;
001 Ron
002 Jan
003 Peter
004 Paul
005 Mary
;
data Grades;
   informat ID $3.;
   input ID Grade1-Grade3;
datalines;
005 78 80 82
002 100 90 95
001 99 95 98
006 65 67 69
004 85 86 84
;
```

The first step in joining these two tables is to select **Query** from the **Utilities** tab.

Figure 4.29: The Query Task in the Utilities Tab

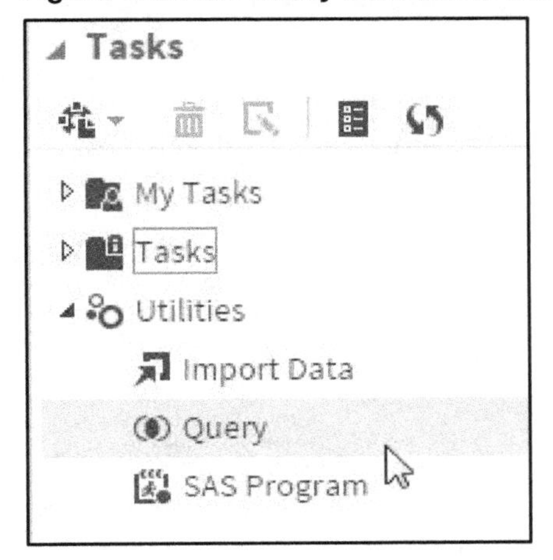

This brings up the following screen:

Figure 4.30: The Query Window

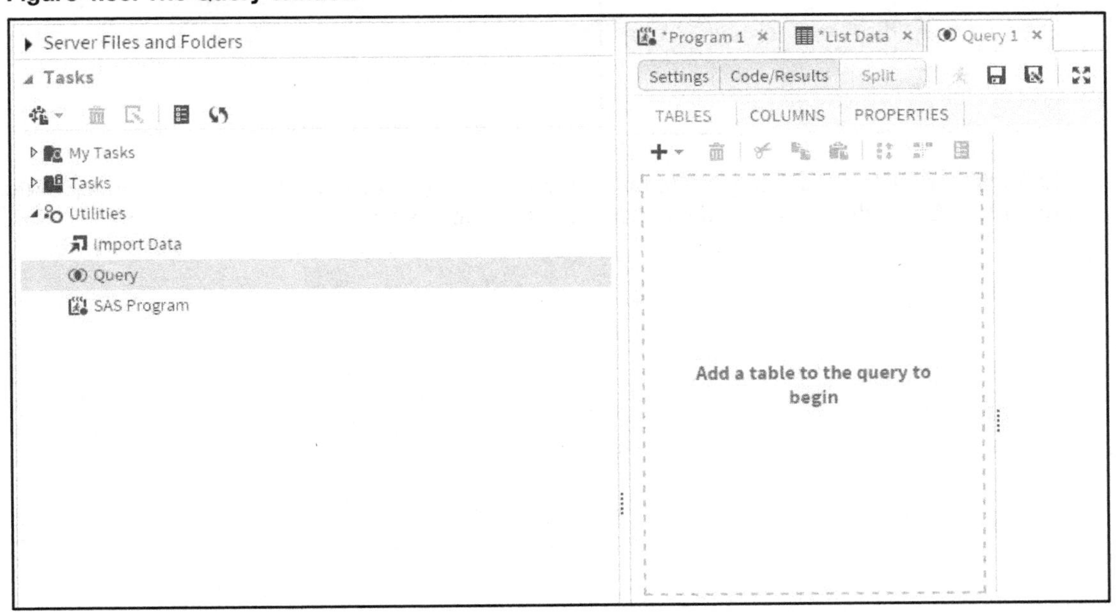

The next step is to open the **Libraries** tab and find the **Work** library.

Figure 4.31: Locate the Two Tables in the Work Library

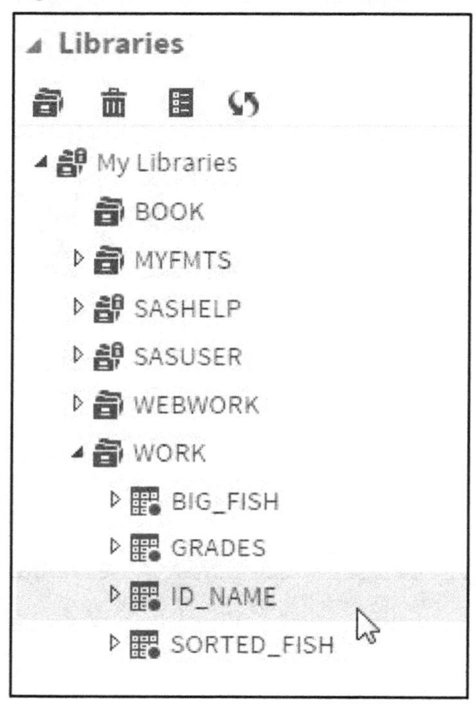

Left-click on each file and drag it into the Query window. (If you drag the second file on top of the first file, SAS Studio automatically assumes that you want to perform a join operation.)

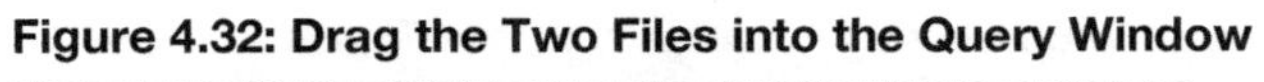

Figure 4.32: Drag the Two Files into the Query Window

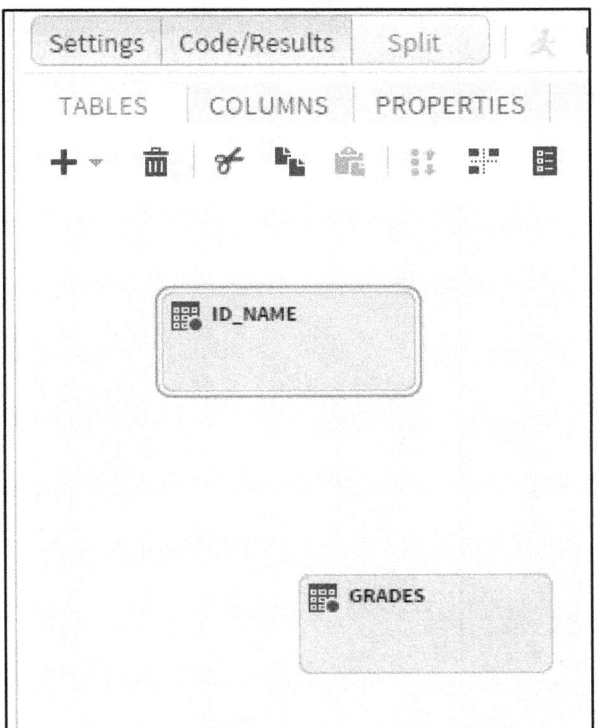

In the pull-down menu on the **TABLES** tab, select **Join** (unless you dragged the second file on top of the first, in which case the Query tool assumes that you want a join).

Figure 4.33: Selecting Join in the Pull-down Menu

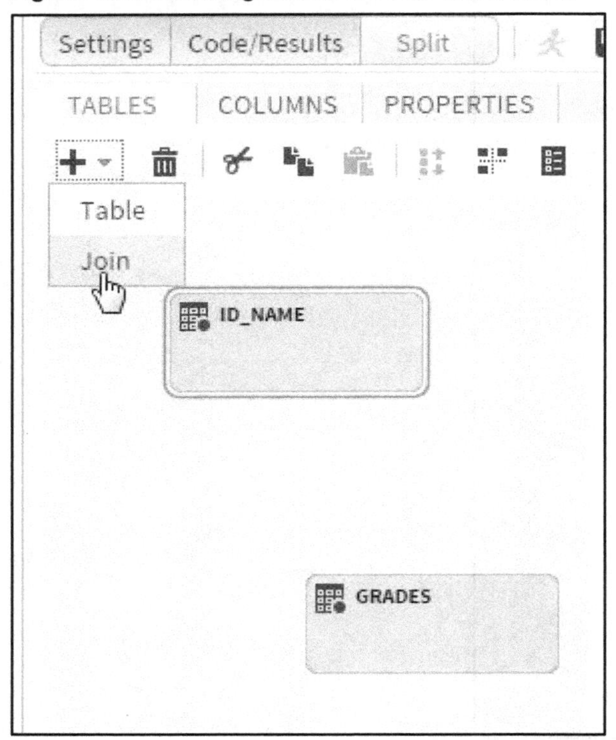

This brings up the following screen:

Figure 4.34: Getting Ready to Join the Two Tables

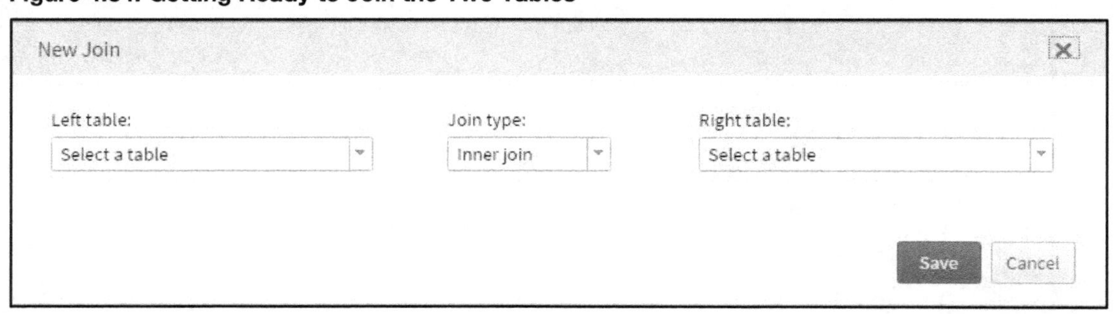

Select **ID_Name** for the **Left table** and **Grades** for the **Right table**. For **Join type**, select **Inner join**.

Figure 4.35: Selecting the Two Tables and Inner Join

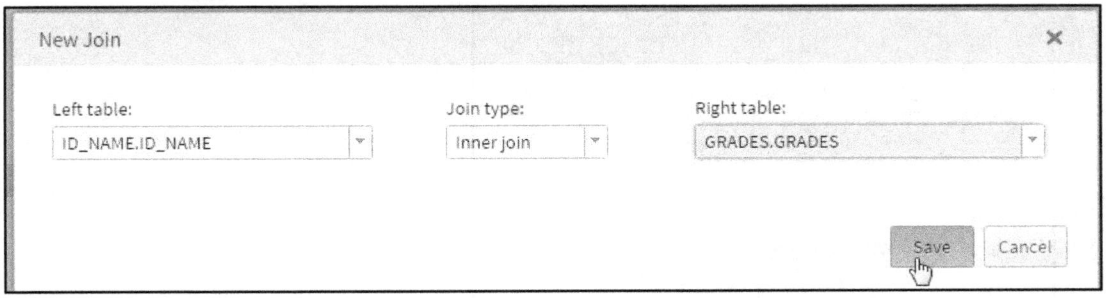

Click **Save**. You now see the two tables with a Venn diagram that represents an inner join. If you are familiar with SQL, you already know the four types of joins. For those readers who are not, here is the explanation:

Because some IDs are only in one table, you have some decisions to make about how you want to handle the join. The most common join, selected in this example, is an *inner join*. This type of join includes only those rows where there is a matching ID in both files. An *outer join* includes all rows from both tables (with missing values in the rows from the table that does not contain an ID). Finally, the other two joins are a *left join* and a *right join*. In a left join, all IDs from the left table are included—in a right join, all IDs from the right table are included.

Figure 4.36: Venn Diagram Showing an Inner Join of the Two Tables

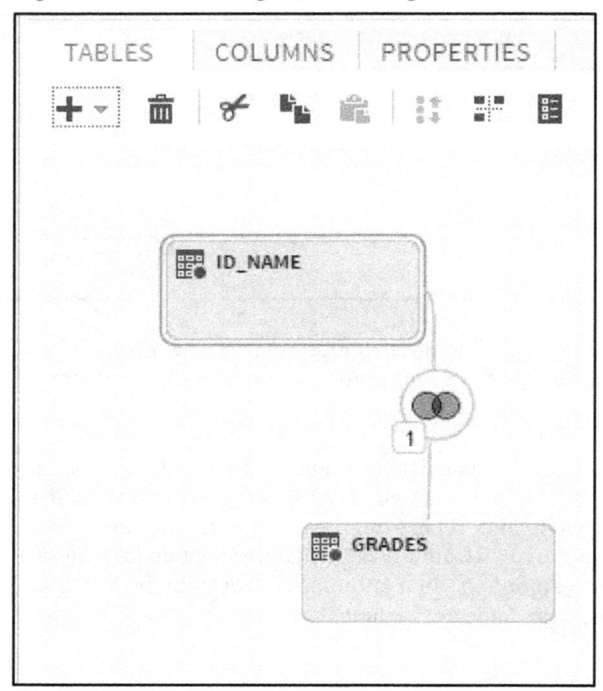

The next step is to name the columns that you want to use to join the tables. In this example, because ID is in both tables, the Query tool automatically selects ID for the join variables. You are free to select any variable from each file to construct the join, even if the variable names are not the same in the two files.

Figure 4.37: Selecting the Join Conditions

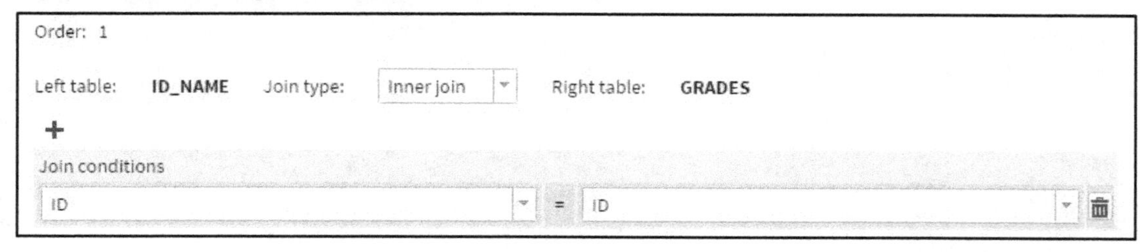

Next, select which columns you want in the joined table. To do this, click on the **COLUMNS** tab. Select the columns in the usual way, and drag them to the Add columns area.

Figure 4.38: Selecting Columns for the Final Table

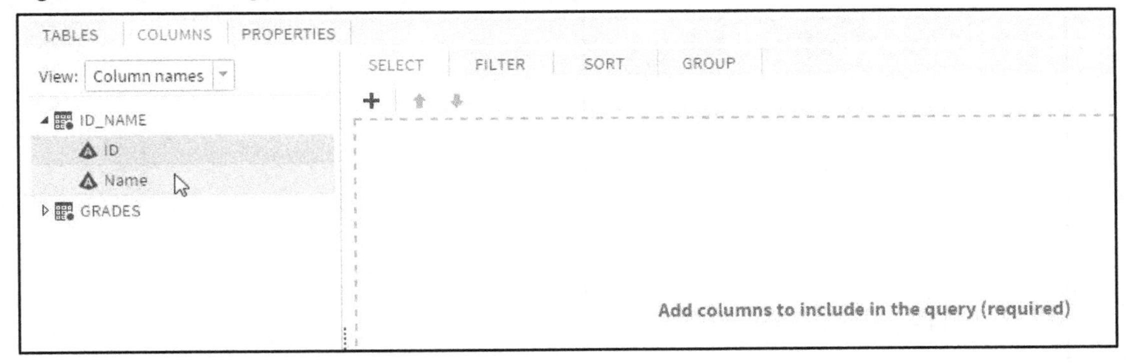

The figure below shows the final list of columns in the joined table.

Figure 4.39: Variables in the Joined Table

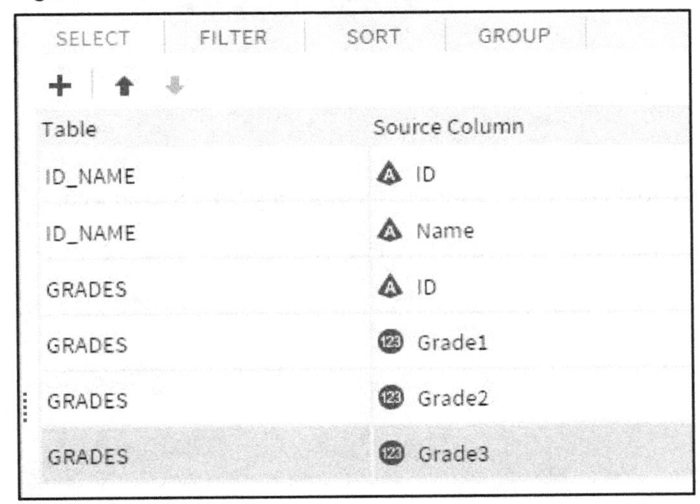

The last step is to click on the **PROPERTIES** tab and indicate if you want a table or a report. If you choose a table, you can name the location (the Work library in this example) and the table name.

Figure 4.40: Options in the PROPERTIES Tab

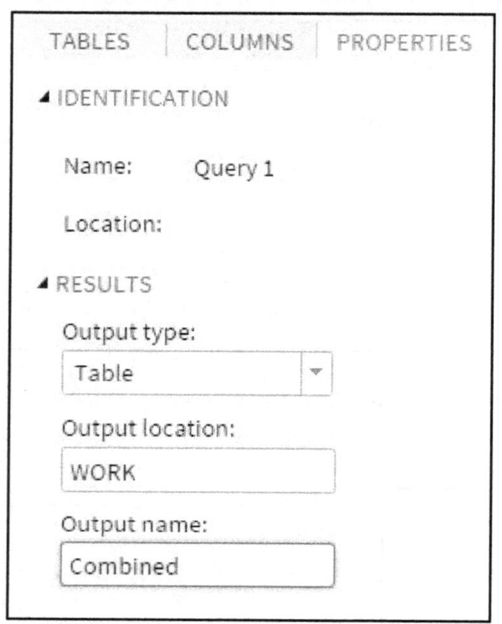

Clicking on the **Run** icon finishes the join. A snapshot view of the resulting table is produced.

Figure 4.41: View of the Resulting Table

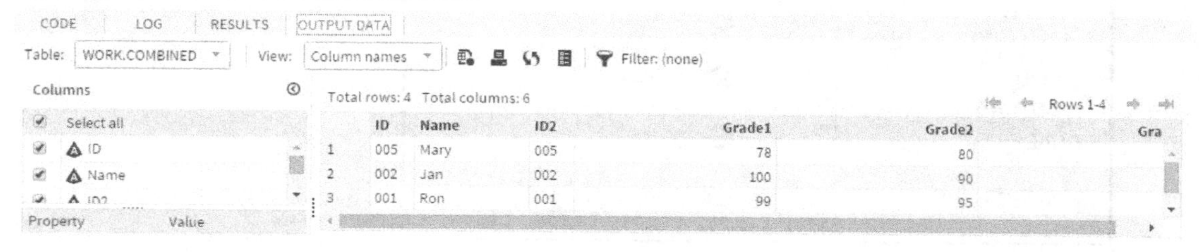

	ID	Name	ID2	Grade1	Grade2	Gra
1	005	Mary	005	78	80	
2	002	Jan	002	100	90	
3	001	Ron	001	99	95	

You can use the **List Data** task to create a listing of the resulting table. It is shown in Figure 4.42:

Figure 4.42: Listing of Combined Data Set

Listing of Data Set COMBINED

ID	Name	ID2	Grade1	Grade2	Grade3
005	Mary	005	78	80	82
002	Jan	002	100	90	95
001	Ron	001	99	95	98
004	Paul	004	85	86	84

Because this was an inner join, only those IDs that were in both tables are listed in the final table.

Conclusion

Only a few of the more popular data tasks were described in this chapter. Once you get the knack of running a task, you should feel confident in trying out some of the other data tasks in the list. The decision to use a task or write a SAS program is a personal choice. For those with programming experience, writing a program may be the way to go—for those folks who are new to SAS and just want to get things done, using the tasks is a great way to go. Or you can do both! Use a task or a utility to get the basic program written for you, and then take that and edit it to do more.

Chapter 5: Summarizing Data Using SAS Studio

Introduction

This chapter demonstrates how to summarize numeric and character data. For numeric variables, you will see how to compute statistics such as means and standard deviations, as well as histograms—for character variables, you will see how to generate frequency distributions and bar charts.

Summarizing Numeric Variables

One of the most useful tasks for summarizing numeric data is found on the **Statistics** task list. Don't be alarmed that this task is listed under statistics—you don't need to be a statistician to understand how this works. Expand the **Statistics** task and select **Summary Statistics**, as shown in Figure 5.1 below:

Figure 5.1: Summary Statistics

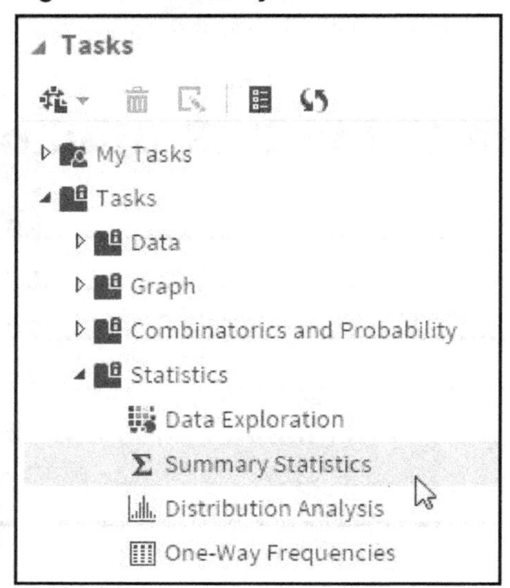

Double-click on **Summary Statistics** to bring up the **DATA** and **OPTIONS** tabs (Figure 5.2).

Figure 5.2: The Summary Statistics Task

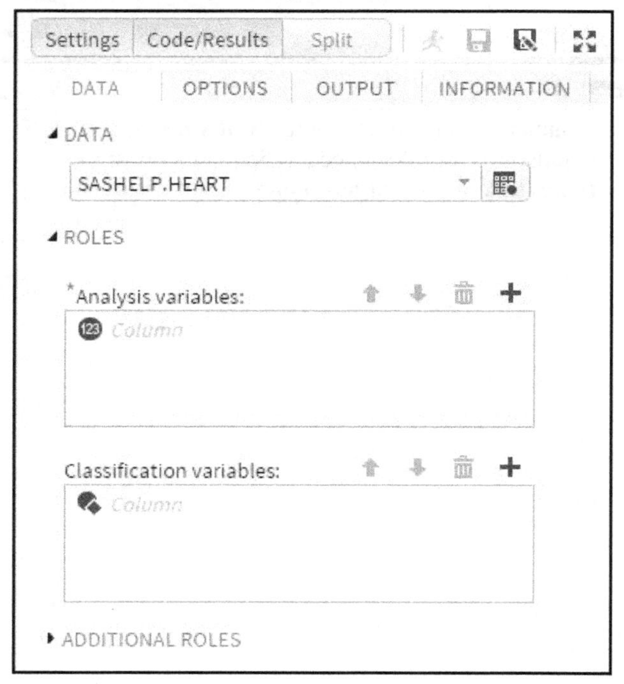

Let's choose the Heart data set in the Sashelp library to demonstrate how to use the **Summary Statistics** task. Looking at Figure 5.2, you see that Sashelp.Heart has already been chosen. You can click on the **Data Table** icon to select a library and data set that you wish to use. The next step is to select the variables you want to summarize.

Figure 5.3: Selecting Variables

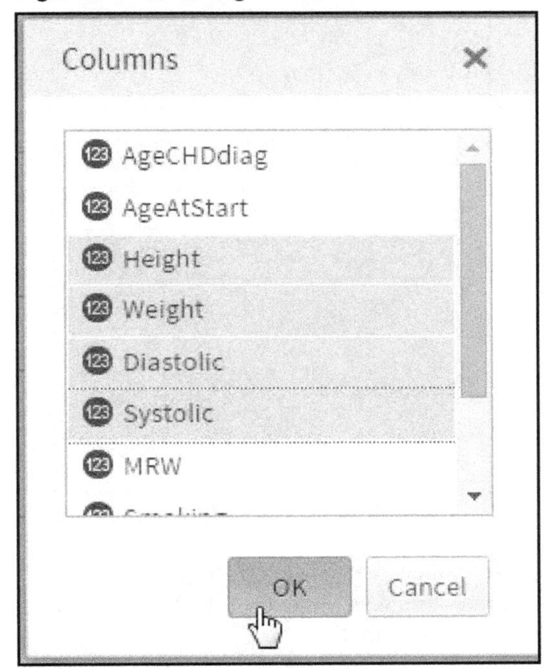

Click on the plus sign (+) in the Analysis Variables window and select the variables you want to analyze (using the Ctrl or Shift keys, as described previously). Notice that the variable list contains only numeric variables. For this example, you have chosen the variables Height, Weight, Diastolic (diastolic blood pressure), and Systolic (systolic blood pressure). Click on the **OK** button when you are done selecting variables.

You can click on the **Run** icon now or customize the report by clicking on the **OPTIONS** tab. Let's do that. It brings up the following screen:

Figure 5.4: Selecting Options

Select or unselect the options you want. This author recommends that you select the two options **Number of observations** and **Number of missing values**—they are quite useful. At this point, you can run the task or continue on to requesting plots. Let's do that.

Figure 5.5: Plots

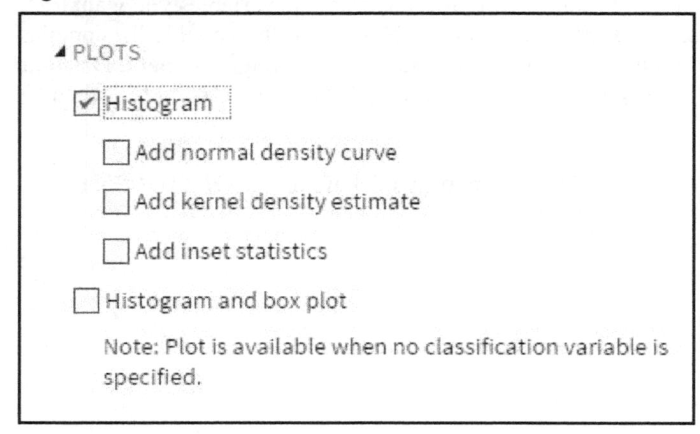

You have the option to include a histogram or a histogram with a box plot. In this example, you have chosen a histogram. If you are statistically minded, you can add a normal density curve and a kernel density estimate. The third option in this list places summary statistics in an inset window in the histogram.

It's time to run the task. The first part of the output shows the statistics you requested in tabular form. It looks like this:

Figure 5.6: Tabular Output

Variable	Mean	Std Dev	Median	N	N Miss
Height	64.8131847	3.5827074	64.5000000	5203	6
Weight	153.0866808	28.9154261	150.0000000	5203	6
Diastolic	85.3586101	12.9730913	84.0000000	5209	0
Systolic	136.9095796	23.7395964	132.0000000	5209	0

Here you see the mean, standard deviation, and median for the selected variables. The last two columns, labeled N and N Miss, show the number of nonmissing observations and the number of missing values, respectively.

To economize on space in this book, only one histogram (Height) is displayed. It looks like this:

Figure 5.7: Histogram

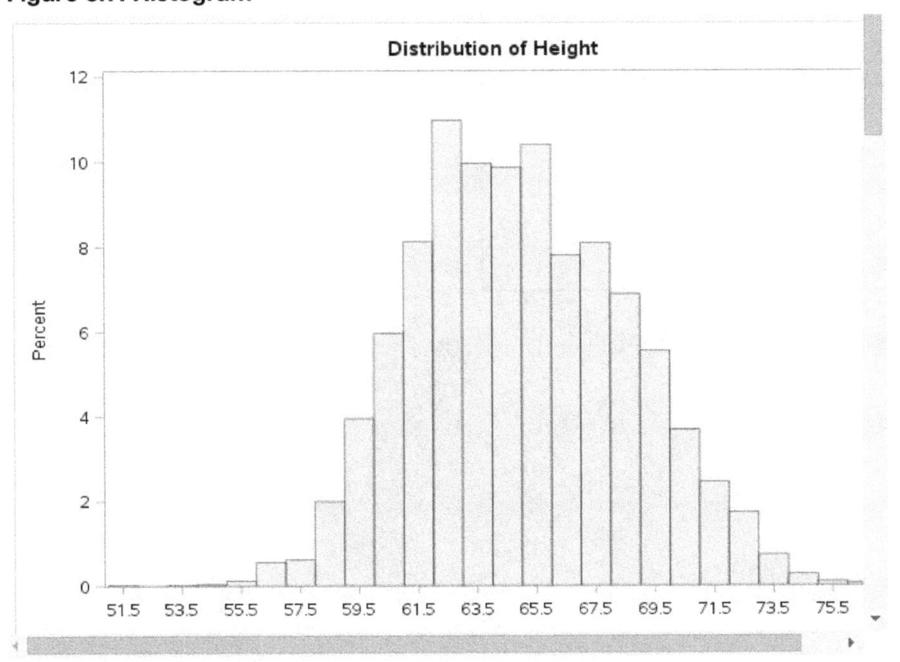

You can use the scroll bars to move right or left, up or down. You can also click on the **Expand** icon to see the entire histogram.

Adding a Classification Variable

The statistics you have seen so far are computed on the entire data set. To see statistics broken down by one or more classification variables, add those variables in the **Classification variables** box. To demonstrate this, let's see the statistics for Height, Weight, Diastolic, and Systolic broken down by the variable Sex. Click on the plus sign in the **Classification variables** box and select the variable **Sex** (Figure 5.8).

Figure 5.8: Adding a Classification Variable

Now run the program to see the following table:

Figure 5.9: Output Showing Classification Data

Sex	N Obs	Variable	Mean	Std Dev	Median	N	N Miss
Female	2873	Height	62.5725863	2.4524112	62.5000000	2869	4
		Weight	141.3886372	26.2880439	138.0000000	2869	4
		Diastolic	84.6463627	13.3394548	82.0000000	2873	0
		Systolic	136.8861817	25.9835883	130.0000000	2873	0
Male	2336	Height	67.5673736	2.7321366	67.5000000	2334	2
		Weight	167.4661525	25.2907044	167.0000000	2334	2
		Diastolic	86.2345890	12.4548941	85.0000000	2336	0
		Systolic	136.9383562	20.6535522	134.5000000	2336	0

You see all of the statistics you originally requested for each value of Sex. If you requested plots, you will see separate histograms for each value of the classification variable. To save space, only the histogram for Height is displayed (Figure 5.10).

Figure 5.10: Separate Histograms by Sex

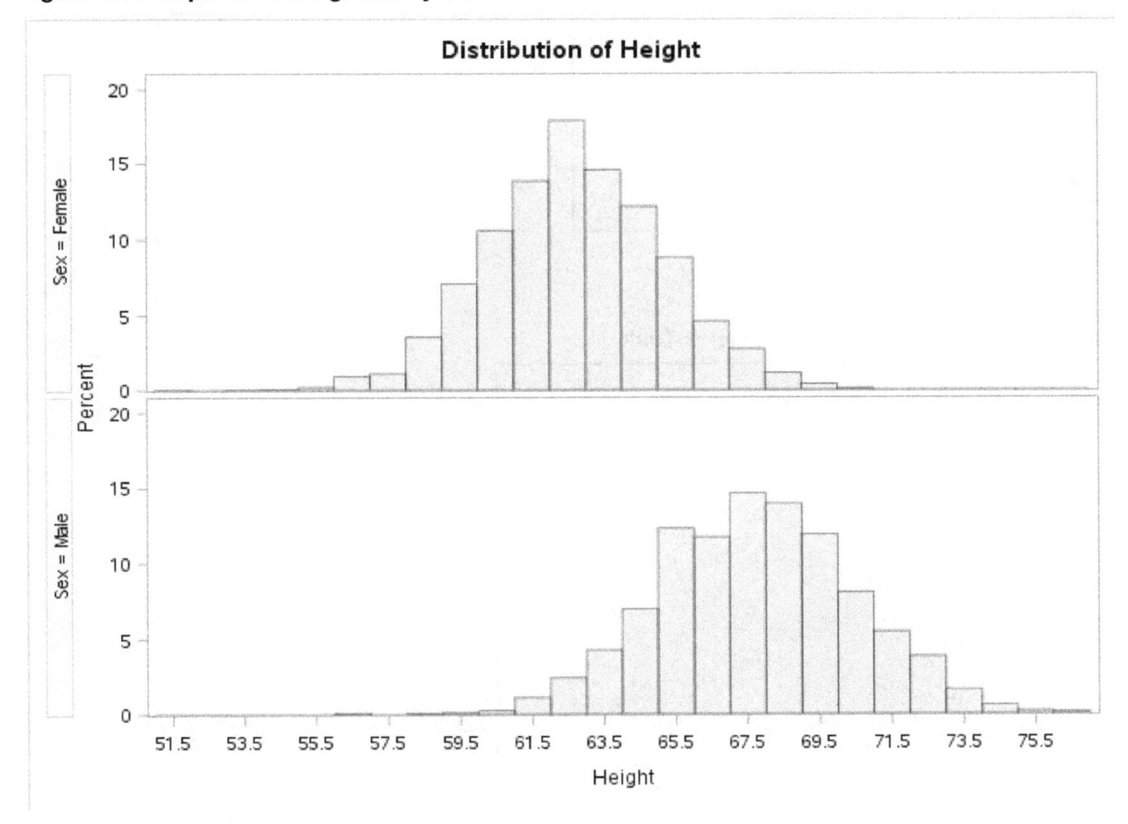

Seeing the two histograms juxtaposed like this is useful in determining if there are differences in the analysis variable for each level of the classification variable.

Summarizing Character Variables

You can use the **One-Way Frequencies** task to compute counts and percentages for character or numeric variables. If you include any numeric variables in your selection, this task will compute frequencies for every unique value of those variables. That is why this task is usually reserved for character variables or for numeric variables with very few unique values.

The first step is to double-click on the **One-Way Frequencies** tab.

Figure 5.11: One-Way Frequencies

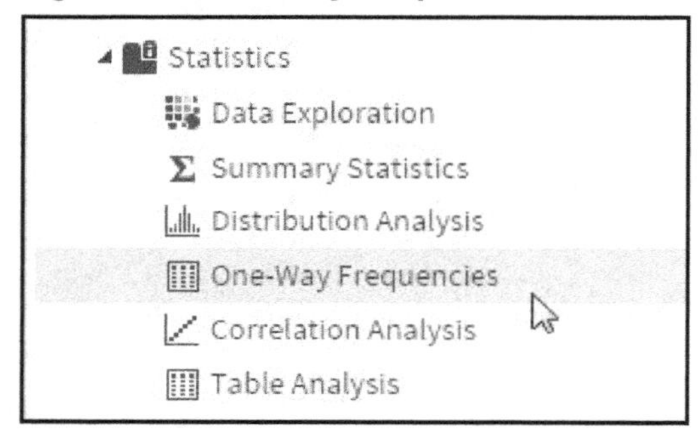

This brings up the screen shown below:

Figure 5.12: One-Way Frequencies Task

The Sashelp.Heart data set has already been selected. Click on the plus sign attached to the **Analysis variables** box. Then select the variables for which you want to compute frequencies. For this demonstration, the variables Sex, Chol_Status, BP_Status, and Smoking_Status were chosen (the variable Sex is farther up the list and does not appear in this screen shot).

Figure 5.13: Selecting Variables

Click the **OK** button to proceed. If you want to customize the frequency table, click on the **OPTIONS** tab. This brings up the following:

Figure 5.14: Frequency Options

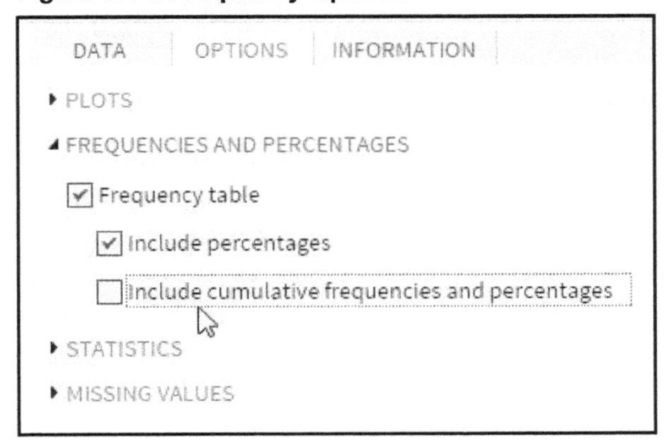

Here, you are unselecting the option to include cumulative frequencies and percentages (the default is to include cumulative frequencies and percentages). If you expand the **PLOTS** option, you will see that the default action is to produce plots. Select the option **Suppress plots** if you do not want bar charts. In this example, the **Suppress plots** option is left unchecked.

Figure 5.15: Suppress Plots

Click on the **Run** icon to complete the task. The output consists of frequency tables and bar charts. Figure 5.16 shows two of the four tables requested. Notice that cumulative frequencies and cumulative percentages are not included (because you unchecked the option to do this).

Figure 5.16: Frequency Tables

Sex	Frequency	Percent
Female	2873	55.15
Male	2336	44.85

Cholesterol Status		
Chol_Status	Frequency	Percent
Borderline	1861	36.80
Desirable	1405	27.78
High	1791	35.42
Frequency Missing = 152		

Only one bar chart (for Smoking_Status) is displayed here (Figure 5.17).

Figure 5.17: Bar Chart

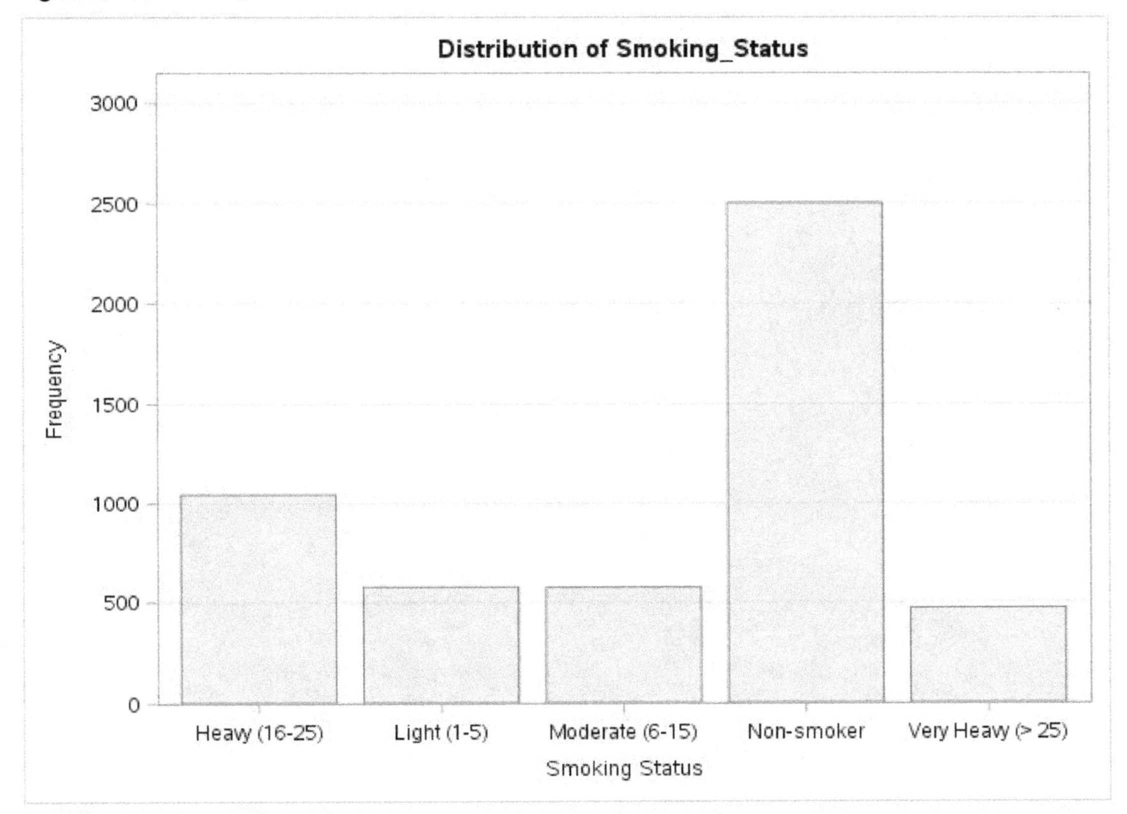

Conclusion

The two statistics tasks, **Summary Statistics** and **One-Way Frequencies**, can be used to summarize numeric and character data, respectively. You can customize the tabular and graphical output by selecting options for both tasks.

Chapter 6: Graphing Data

Introduction

Creating charts and graphs is one of the more difficult programming tasks. Even veteran programmers need to pull out the manual or seek online help when attempting these tasks. Luckily, SAS Studio has a number of charting and graphing tasks that allow you to create beautiful, customized charts and graphs with ease.

This chapter will demonstrate a few of the more popular tasks, such as creating bar charts, pie charts, and scatter plots. Once you see how these tasks work, you will be able to use any of the other graph tasks offered by SAS Studio.

Creating a Frequency Bar Chart

Let's start out with a simple bar chart where each bar represents a frequency on the y-axis. Because you are somewhat familiar with the Sashelp data set Heart (which is used in many of the previous chapters), let's start out using the variable Smoking_Status to create a bar chart.

Open the **Tasks** tab in the navigation pane and expand the **Graph** tasks. It looks like this:

Figure 6.1: Graph Tasks

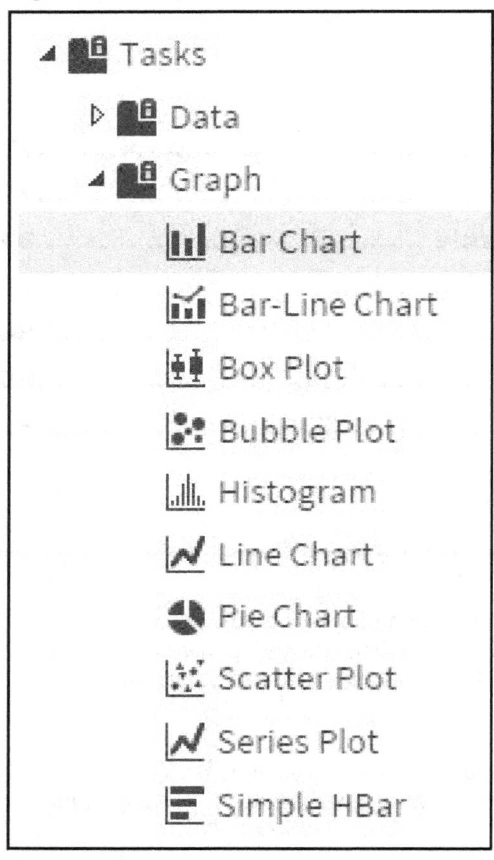

Double-click **Bar Chart** to get started. This opens the following screen:

Figure 6.2: Requesting a Bar Chart

The Heart data set has already been entered and **Smoking_Status** was chosen as the **Category variable**. The next step is to select options. Click on the **OPTIONS** tab to do this.

Figure 6.3: Bar Chart Options

DATA OPTIONS INFORMATION

▲ TITLE AND FOOTNOTE

Title:

Bar Chart Example

☐ Set title font size

*Font size: 14

Footnote:

☐ Set footnote font size

*Font size: 12

▸ BAR DETAILS

▸ BAR LABELS

▸ CATEGORY AXIS

▸ RESPONSE AXIS

▸ LEGEND DETAILS

▸ GRAPH SIZE

You can enter titles and footnotes for this task and even choose the font size for each one. As you can see in Figure 6.3, you can also customize the bar details (change the bar color, for example), the bar labels, the two axes, the chart legend, and the graph size. You can also decide to accept all the defaults for these options, and click on the **Run** icon. You can always go back and tweak these options later if you want. The figure below shows the resulting bar chart (accepting all the defaults):

Figure 6.4: Final Bar Chart

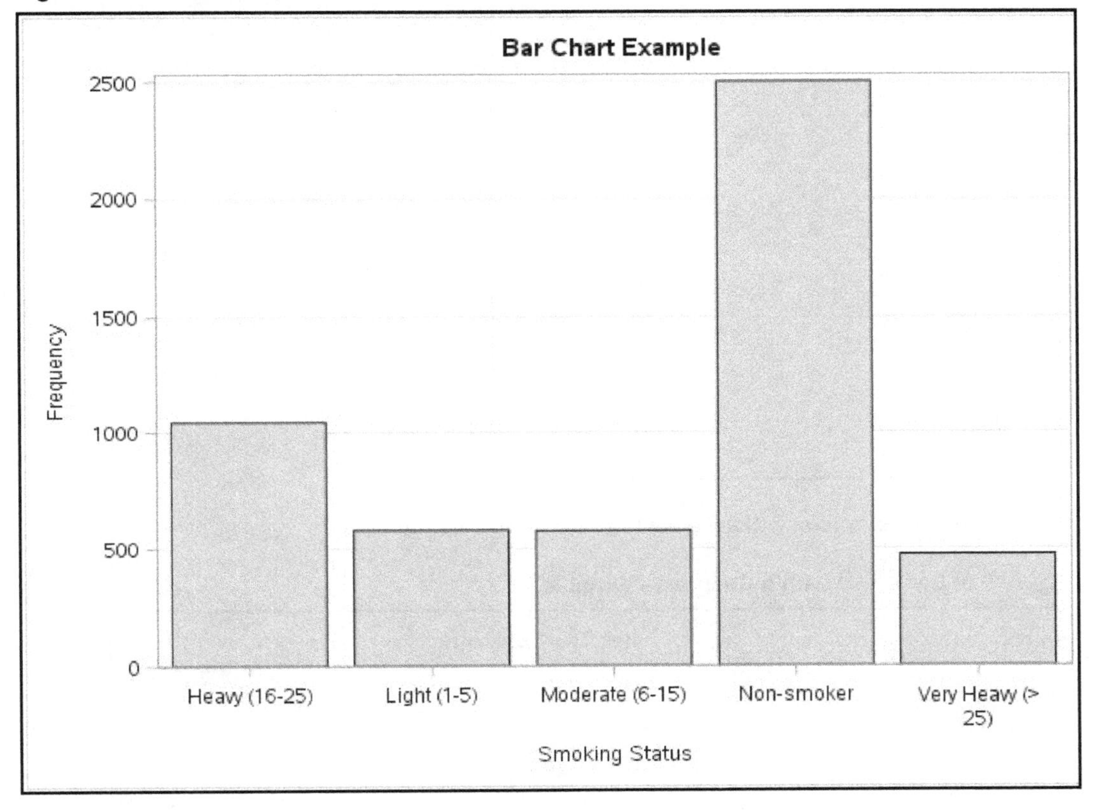

Creating a Bar Chart with a Response Variable

The height of each bar in the previous chart represented frequency counts for each category of Smoking_Status. By choosing a response variable, you can create a bar chart where the height of each bar represents a statistic (the mean, for example) for a response variable at each level of smoking status.

Figure 6.5 shows the **DATA** tab for the **Bar Chart** task with **Weight** entered as the response variable. Because the default statistic is the mean (average), the height of each bar will now represent the mean weight for each value of smoking status. There are options to change the response statistic to a sum rather than a mean, to orient the bars horizontally, and to stack or cluster the bars (only for frequency charts with a group variable).

Figure 6.5: Entering a Response Variable

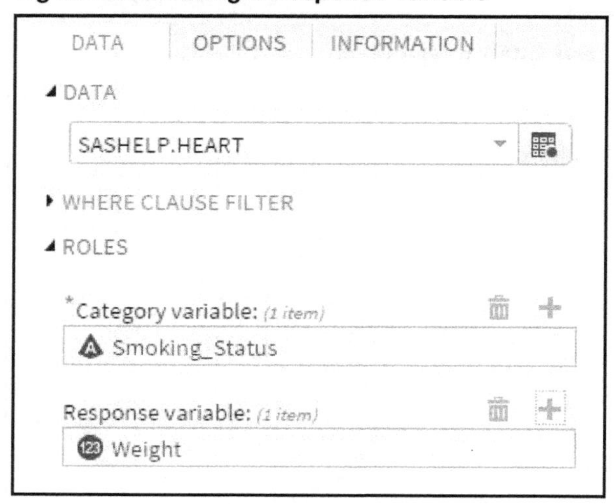

Clicking on the **Run** icon will generate the chart shown in Figure 6.6:

Figure 6.6: Bar Chart with a Response Variable

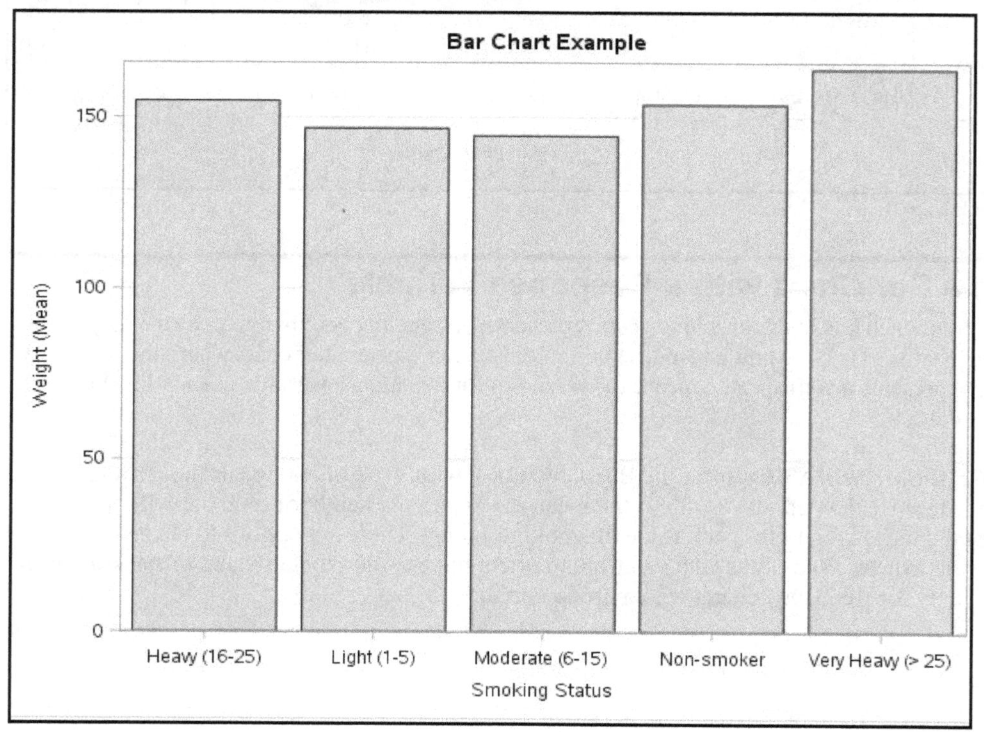

The height of each bar represents the mean weight for each category of smoking status.

Adding a Group Variable

You can display more information in the bar chart by including a group variable in the **DATA** tab. Figure 6.7 shows that you want the variable Sex as a group variable. At this point, you could choose to create a stacked chart (the male and female frequencies in a single bar with two different colors) or the default chart that shows side-by-side bars for men and women.

Figure 6.7: Adding a Group Variable

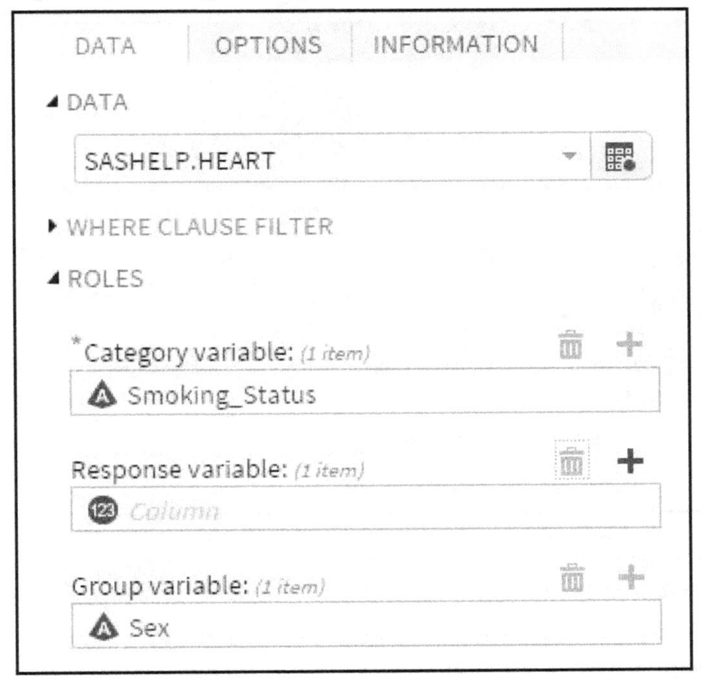

If you do not want to enter any more options, run the task to create the cluster bar chart in Figure 6.8:

Figure 6.8: Grouped Bar Chart

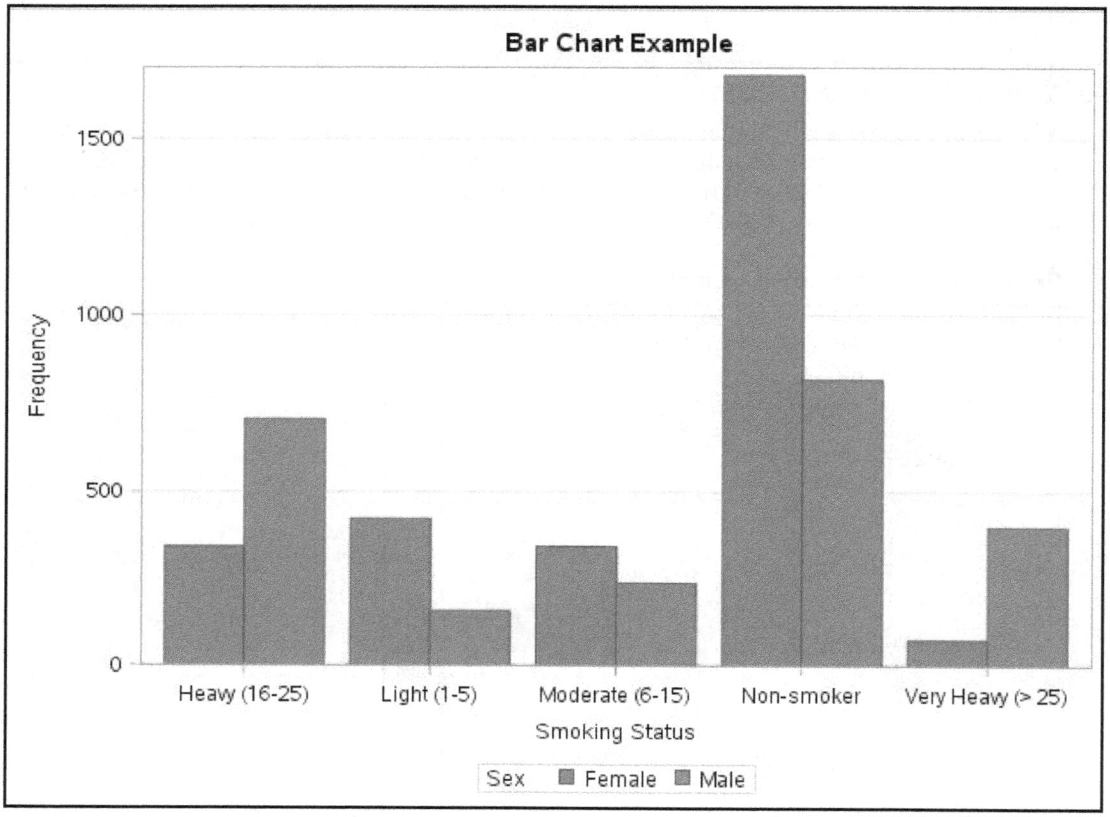

It looks like there were many more female non-smokers in this sample, while in the heavy and very heavy categories, males predominated.

Creating a Pie Chart

Pie charts represent a popular way to display frequencies. As with bar charts, the size of the slices can also represent a sum or mean of a response variable. Let's start out by generating a pie chart showing smoking status. Select **Pie Chart** from the task list.

Figure 6.9: Requesting a Pie Chart

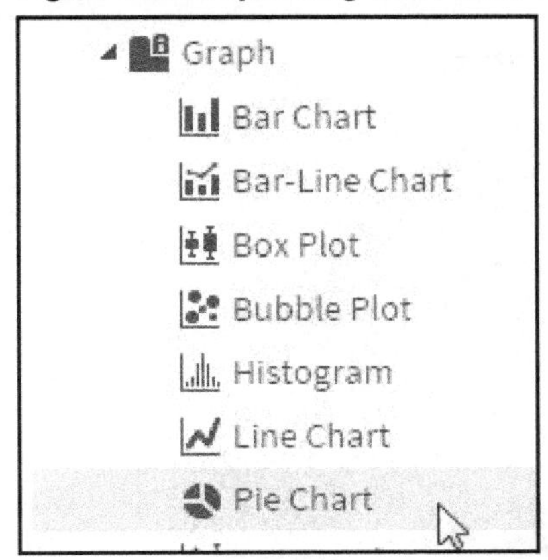

Double-click on **Pie Chart** to get started. This brings up the following screen:

Figure 6.10: Choosing a Variable for the Pie Chart

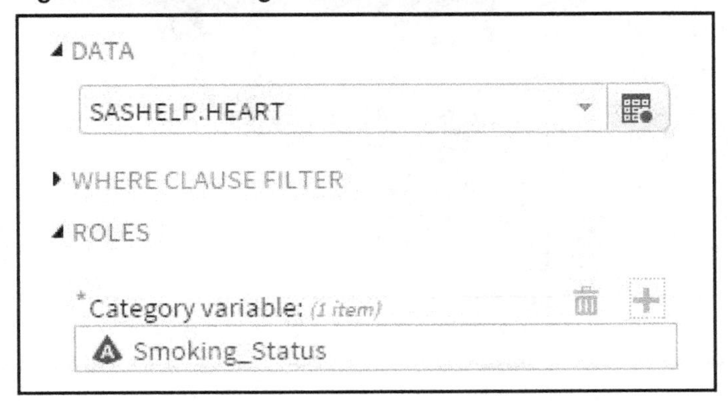

There are a number of options that control color, transparency, and visual effects for the pie slices. You can modify them or accept the defaults and create the chart. A chart with all the default options is shown in Figure 6.11.

Figure 6.11: Final Pie Chart

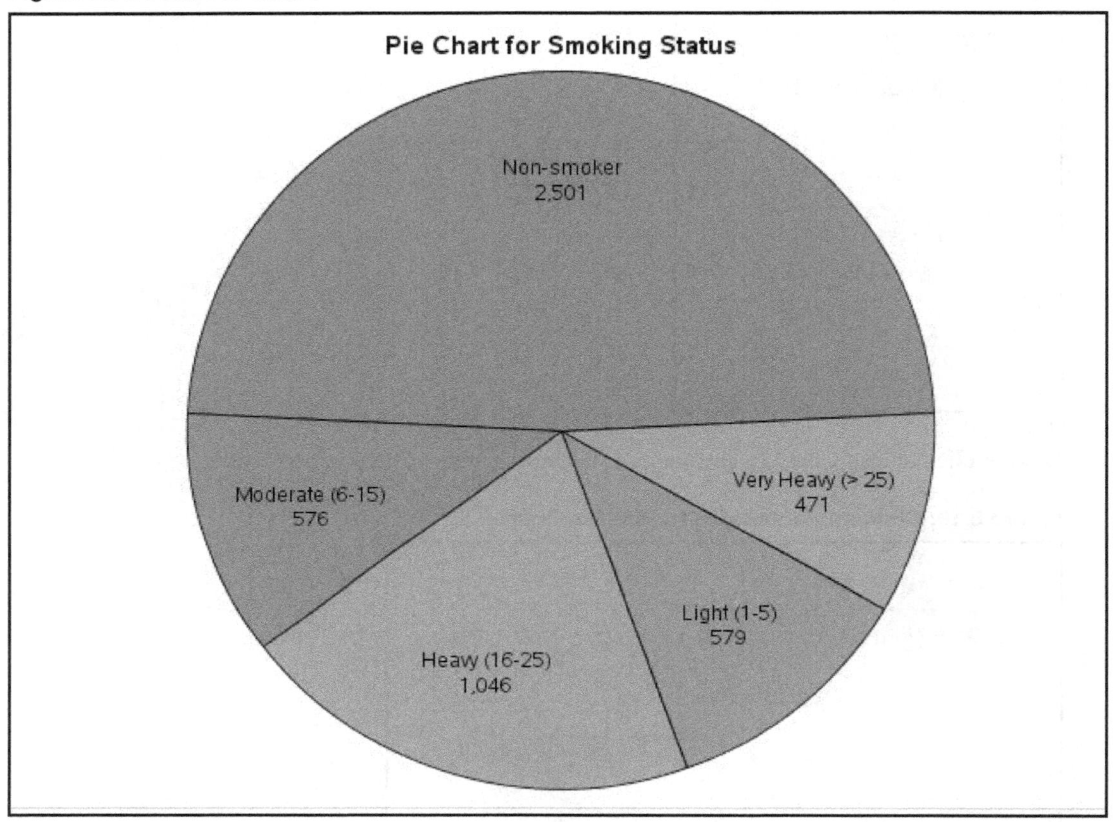

Creating a Scatter Plot

A scatter plot shows the relationship between two variables by placing points on a set of x- and y-axes. To get started, locate **Scatter Plot** in the **Graph** task list.

Figure 6.12: Requesting a Scatter Plot

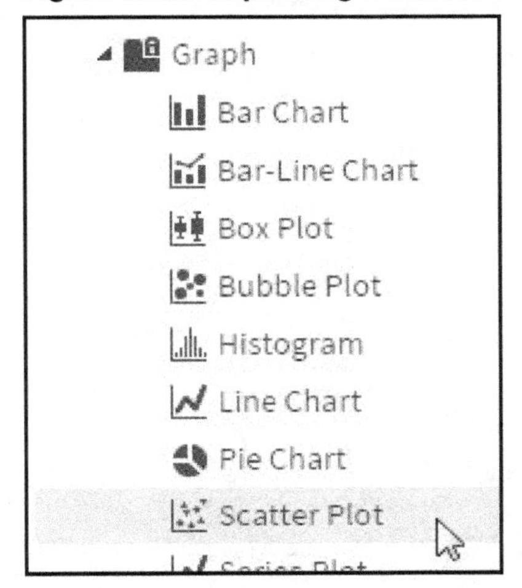

Double-click to bring up the following screen:

Figure 6.13: Selecting the X, Y, and Grouping Variables

You enter the x- and y-variables as shown. Here you want to see Height on the x-axis and Weight on the y-axis. You can also identify a group variable. For this example, you are choosing Sex as a group variable. Click the **Run** icon to generate the really impressive scatter plot shown in Figure 6.14:

Figure 6.14: Final Scatter Plot

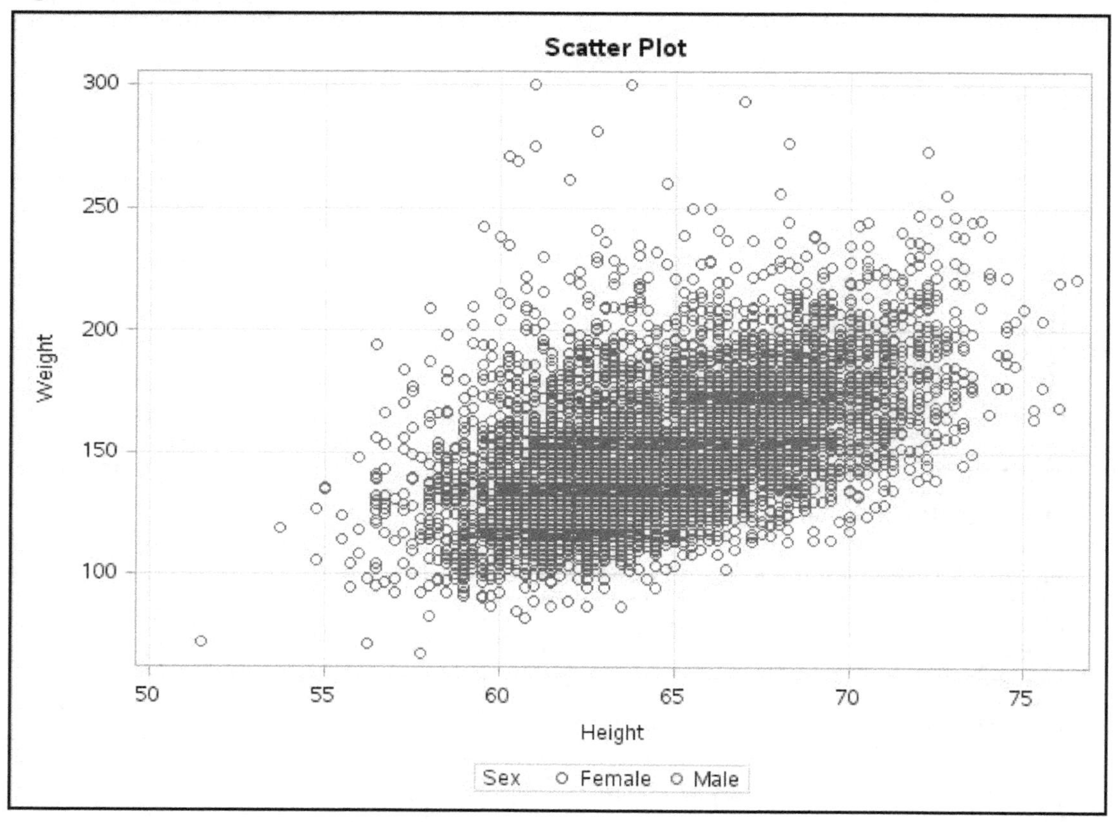

It appears that there is a positive relationship between Height and Weight. Also, the taller and heaver data is male-dominated.

Conclusion

After seeing how to run the charts and graphs demonstrated in this chapter, you should have no trouble running any of the graph tasks included with SAS Studio. This author, for one, is immensely grateful that someone has already done all the behind-the-scenes work to make it so easy to create charts and graphs. Thank you, SAS!

Part 2: Learning How to Write Your Own SAS Programs

Part II shows you how to write your own SAS programs and how to use SAS procedures to perform a variety of tasks. This section also explains how to read data from a variety of sources, including text files, Excel workbooks, and CSV files. In order to help you become familiar with the SAS Studio environment, the book also shows you how to access the dozens of interesting data sets that are included with the product.

Chapter 7: An Introduction to SAS Programming

SAS as a Programming Language

This section of the book is dedicated to teaching you how to write your own programs in SAS. Perhaps you have some programming experience with other languages, such as C+, Python, or Java. This is both an advantage and a possible disadvantage. The advantage is that you understand how to think logically and use conditional logic, such as IF-THEN-ELSE statements and DO loops. On the other hand, SAS is somewhat unique in the way it reads and processes data, so you need to "re-wire" your brain and start to think like a SAS programmer.

SAS programs consist of DATA steps, where you read, write, and manipulate data and PROC (short for procedure) steps, where you use built-in procedures to accomplish tasks such as writing reports, summarizing data, or creating graphical output. DATA steps begin with the keyword DATA and usually end with a RUN statement. PROC steps begin with the word PROC (did you guess that?) and end with either a RUN or QUIT statement (or both).

> SAS statements all end with semicolons. This is a good place to mention that one of the most common programming mistakes, especially with beginning SAS programmers, is to forget a semicolon at the end of a statement. This sometimes leads to confusing error messages.

The SAS Studio Programming Windows

When you open up SAS Studio, you see three tabs: **CODE**, **LOG**, and **RESULTS** (see Figure 7.1 below).

Figure 7.1: The Three SAS Studio Windows

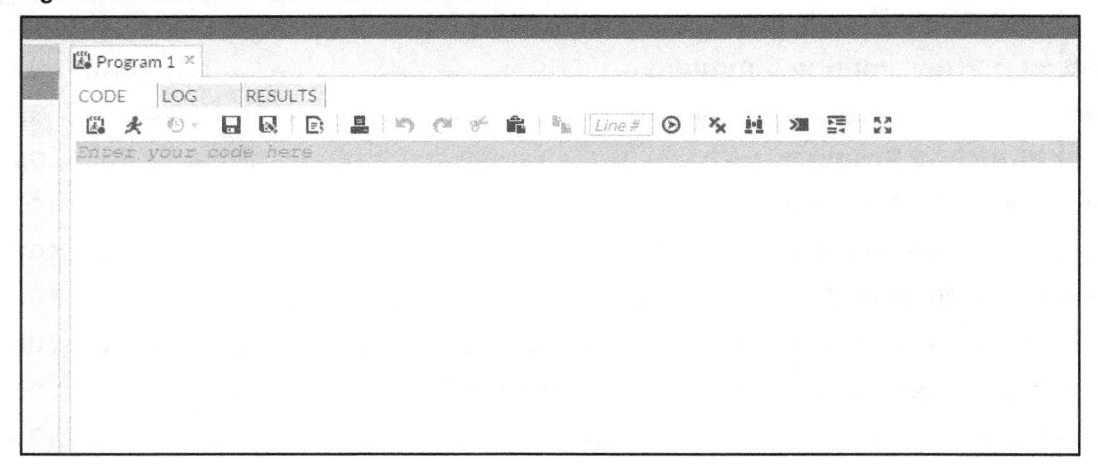

The CODE window is where you write your SAS program. When you run a SAS program, the LOG window displays your program, any syntax errors detected by SAS, information on data that was read or written out, and information about real time and CPU time used. The RESULTS window is where any SAS output appears. You can navigate among the three windows by clicking on the appropriate tab.

Your First SAS Program

Let's first write a simple program and then follow what happens when it runs. Suppose you want a program that converts temperatures from Celsius to Fahrenheit. It is a good idea to start your program by writing a comment statement. As part of the comment, you should, at a minimum, state the purpose of the program. In a more formal setting, you might also include information such as who wrote the program, the date it was written, and the location of input and/or output files. One way of writing comments in SAS programs is to start the comment with an asterisk and end it with a semicolon.

The next line is a DATA statement where you give a name to the data set you are going to create. Look what happens as you start to write the word DATA:

Figure 7.2: Illustrating the Autocomplete Feature of SAS Studio

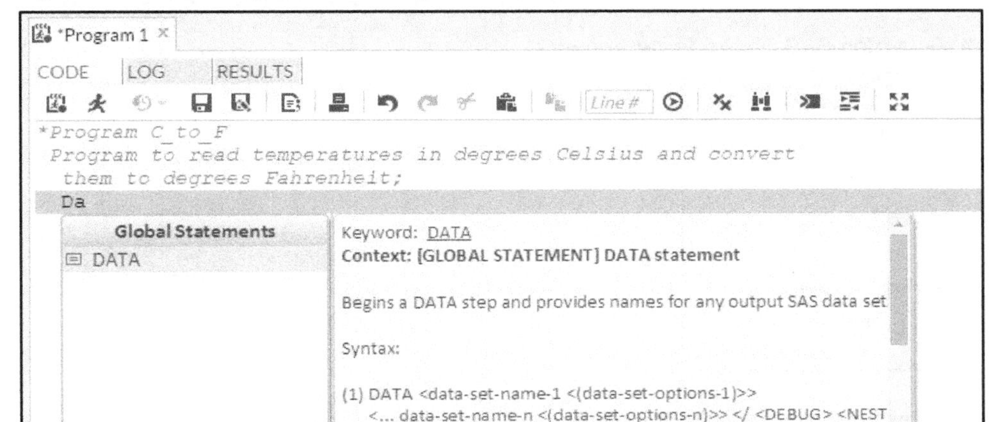

As you type certain keywords in the CODE window, context-sensitive boxes pop up to show you syntax and options that are available for the statement you are writing. If you are a more advanced user who does not need this syntax help, you can turn it off by clicking on the menu icon (on the top right of the screen) as follows:

Figure 7.3: SAS Studio Options

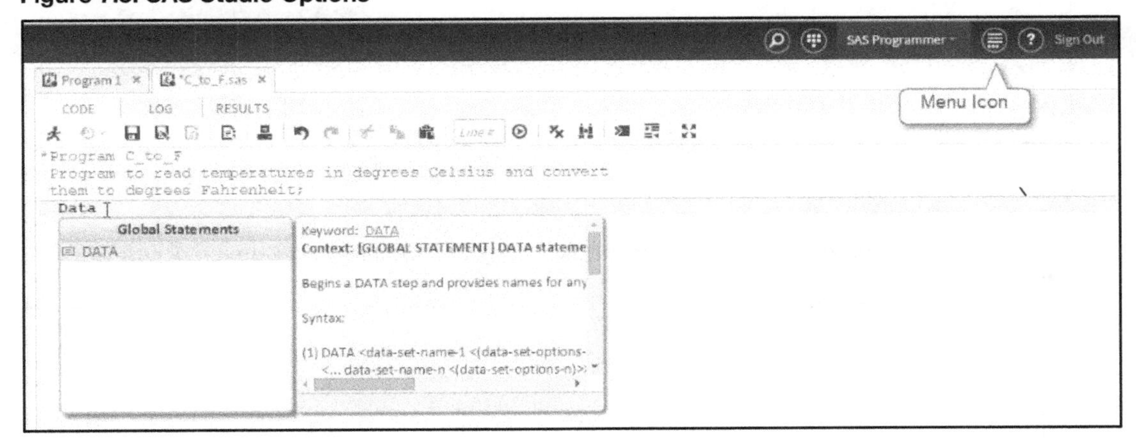

Choose the **Preferences** tab on this menu:

Figure 7.4: Selecting SAS Studio Preferences

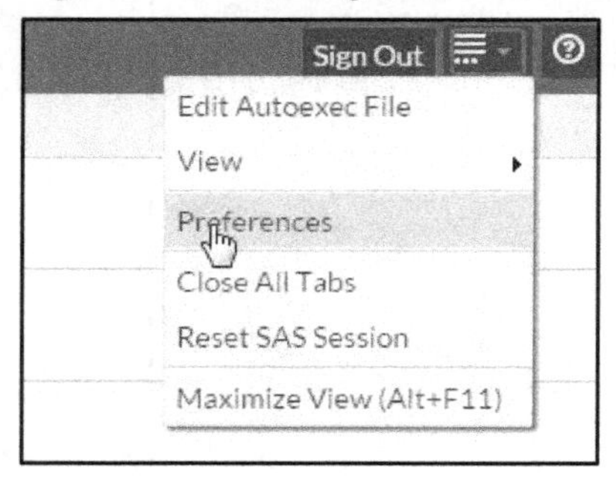

Under the **Editor** tab, select (or deselect) **Enable autocomplete**.

Figure 7.5: Select or Deselect Autocomplete

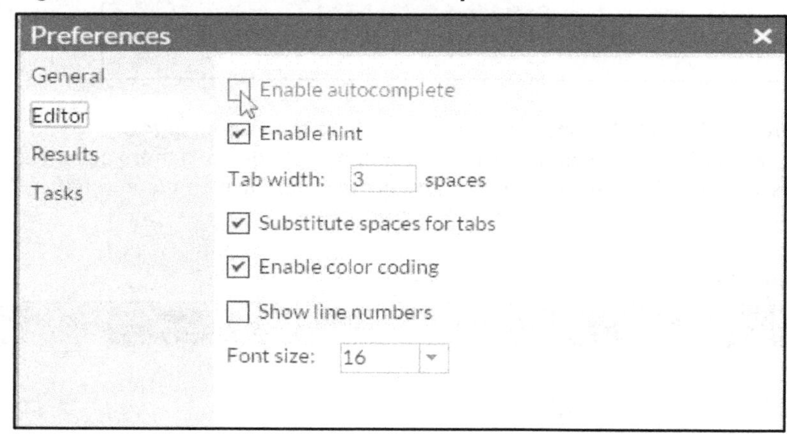

The complete program is listed below:

Program 7.1: Program to Read Temperatures in Degrees Celsius and Convert Them to Fahrenheit

```
*Program C_to_F
 Program to read temperatures in degrees Celsius and convert
 them to degrees Fahrenheit;
Data Convert;
   infile "/folders/myfolders/celsius.txt";
   input Temp_C;
   Temp_F = 1.8*Temp_C + 32;
run;
```

```
title "Temperature Conversion Chart";
proc print data=Convert;
   var Temp_C Temp_F;
run;
```

In this program you name the data set Convert. The rules for naming data sets and many other SAS names (such as variable names) are as follows:

> In SAS, data set names and variable names must start with a letter or underscore. They may contain a maximum of 32 characters, and the remaining characters must be letters, digits, or underscores.

The following tables show examples of valid and invalid SAS names:

Valid SAS Names
My_Data
HeightWeight
X123
_123
Price_per_pound

Invalid SAS Names	
My Data	Contains an invalid character (space)
123xyz	Starts with a digit
Temperature-Data	Contains an invalid character (-)
Group%	Contains an invalid character (%)

> In SAS, variable names are not case-sensitive. However, the case you use the first time you reference a variable is used in SAS output, regardless of how you write the variable name in other locations in the program.

The INFILE statement tells the program where the raw data is located (use a different statement if you have cooked data). Your virtual machine is running a Linux operating system where naming conventions for files are different from the naming conventions used on Microsoft or Apple computers. Filenames in Linux are case-sensitive and folders and subfolders are separated by forward slashes. Filenames on Microsoft platforms are not case-sensitive and folders and subfolders are separated by backward slashes. To help resolve these file naming conventions, you set up shared folders on your virtual machine that allow your SAS programs to read and write files to the hard drive on your computer.

There are slight differences in how you create shared folders, depending on whether you are running VirtualBox, VMware Player, or VMware Fusion. The easiest way to read and write data between your SAS Studio session and your hard drive is to place your data files in a specific location—`\SASUniversityEdition\myfolders`. This location is mapped to a shared folder called `/folders/myfolders` in SAS Studio if you follow the installation directions.

For most of the examples in this book, the location `c:\SASUniversityEdition\myfolders` is the folder where your data files and SAS data sets are located. You can place `\SASUniversityEdition\myfolders` in My Documents or some other location of your choosing. All the programs and data files that you place in `\SASUniversityEdition\myfolders` will show up when you click the **SERVER FILES AND FOLDERS** tab in SAS Studio. Later on, you will see how to set up other shared folders that allow you to read files from any location on your computer.

The raw data file Celsius.txt is stored in the location: `c:\SASUniversityEdition\myfolders\Celsius.txt`. To access this file, you use the location `/folders/myfolders/Celsius.txt` on your INFILE statement. If you are unsure where a file is located, you can right-click on the file in the folder shortcuts and select **PREFERENCES**. This will display the location you need to use on your INFILE statement. You can copy this location (Control-C) and paste it in your program (Control-V). SAS Studio automatically translates this location to the actual location on your hard drive. This is probably the most complicated aspect of running an application on a virtual machine. However, once you do this a few times, it will get easier.

If you want to run the programs in this book, remember that you can download all of the programs and data sets from the following location:

```
support.sas.com/cody
```

Once you arrive at this location, click on the link "Example Code and Data" for this book. This action will download a ZIP file containing all the programs and data files in the book to your hard drive. You will want to extract these files to the `myfolders` folder on your computer.

The INPUT statement is an instruction to read data from the Celsius.txt file. This text file contains one number per line and is listed below:

File Celsius.txt

```
0
100
20
```

This INPUT statement uses one of three methods of reading data, called *list input*. When you use list input, you can read data values separated by blanks (the SAS default delimiter) or other delimiters such as commas. Information on how to process data with delimiters other than blanks is presented in the next chapter.

Following the INPUT statement, you write the formula for the Celsius to Fahrenheit conversion. As with most programming languages, you use an asterisk to indicate multiplication, a forward slash for division, plus and minus signs for addition and subtraction, and two asterisks for exponentiation. Exponentiation is performed first, multiplication and division next, followed by addition or subtraction. You can always use parenthesis to determine the order of operation.

The DATA step ends with a RUN statement. In this program, the statements in the DATA step and PROC step are indented. This is not necessary, but it makes the program easier to read. It is also possible to place more than one SAS statement on a single line, as long as each statement ends with a semicolon; however, this practice is discouraged because it makes it difficult to read and understand the program. Finally, a SAS statement may use as many lines as necessary, such as the comment statement in this program. Just remember to end the statement with a semicolon.

SAS Studio has an auto formatting feature that you can use to automatically format your SAS programs. After you have written your program in the CODE window, click on the **AUTOFORMAT** icon at the top of the Editor window (as shown in the figure below):

Figure 7.6: Activating the Auto Formatting Feature of SAS Studio

The result is a nicely formatted program. You can use Control-Z to undo the formatting in case you don't like the result.

How the DATA Step Works

Unlike most programming languages, the SAS DATA step is actually an automatic loop: The first time the INPUT statement executes, the program reads a value from the first line of data. It then computes the corresponding Fahrenheit temperature according to your formula. Your next statement is a RUN statement that marks the end of the DATA step. Two things happen at this point: First, the program automatically outputs an observation (containing the variables Temp_C and Temp_F) to the output data set that you named Convert. Next, the program performs its implied loop by returning to the top of the DATA step to execute the INPUT statement again. On the second iteration of the DATA step, SAS reads data from the second line of data. Each time the DATA step iterates, the INPUT statement goes to a new line of data (unless you give it special instructions not to). In this example, the DATA step stops when it tries to read the fourth line of data from the file and encounters an end-of-file marker. At this point, your SAS data set Convert contains three observations.

You use PROC PRINT to list the contents of your SAS data set. In this example, you specify the name of the SAS data set using the procedure option DATA= . You can specify one or more title lines with a TITLE statement. You can place a TITLE statement before or after the PROC statement. When you specify a title, that title will remain in effect until you replace it with another TITLE statement. In this program, you placed the TITLE statement between the DATA and PROC steps, a location referred to as *open code.*

You use a VAR statement to specify which variables you want to include in your report. The order you use to list the variables on the VAR statement is the order that PROC PRINT will use to print the results. If you leave out a VAR statement, PROC PRINT will print every variable in your data set in the order they are stored in the data set. You end the PROC step with a RUN statement.

You are now ready to run the program. You do this by clicking on the **RUN** icon as shown in Figure 7.7:

Figure 7.7: The RUN Icon

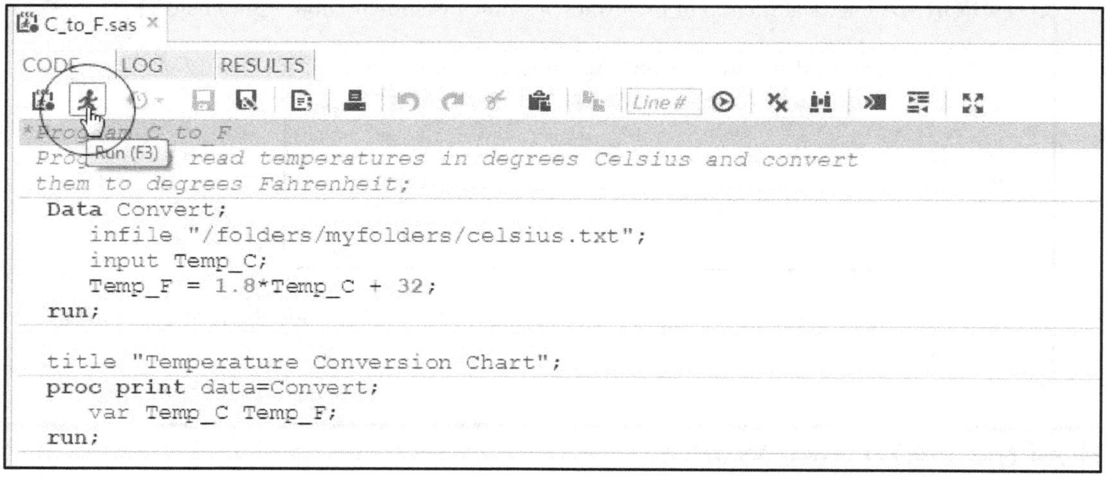

C_to_F.sas

CODE LOG RESULTS

Run (F3)

```
*Program C_to_F
Program to read temperatures in degrees Celsius and convert
them to degrees Fahrenheit;
 Data Convert;
    infile "/folders/myfolders/celsius.txt";
    input Temp_C;
    Temp_F = 1.8*Temp_C + 32;
 run;

 title "Temperature Conversion Chart";
 proc print data=Convert;
    var Temp_C Temp_F;
 run;
```

SAS Studio now shows you the results:

Figure 7.8: The RESULTS Window

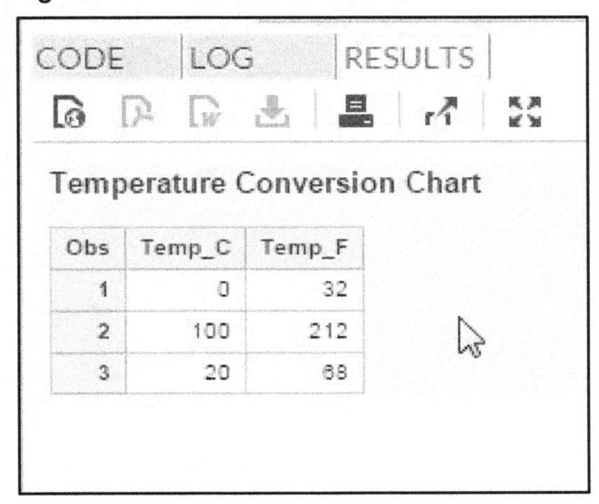

Obs	Temp_C	Temp_F
1	0	32
2	100	212
3	20	68

Be default, SAS lists all rows (called *observations* in SAS terminology) and all columns (called *variables* by SAS). It also includes an Obs. (observation number) column. A quick examination of the output shows that the program worked correctly. However, **you should always look at the SAS log**, even when you have output that seems to be correct. To examine the log, click on the **LOG** tab. Below is a listing of the log:

Figure 7.9: The LOG Window

The first part of the log shows there were no errors or warnings. You can click on any of these items to display any errors, warnings, or notes.

```
1          OPTIONS NONOTES NOSTIMER NOSOURCE NOSYNTAXCHECK;
38         ;
39         *Program C_to_F
40          Program to read temperatures in degrees Celsius and convert
41          them to degrees Fahrenheit;
42           Data Convert;
43              infile "/folders/myfolders/celsius.txt";
44              input Temp_C;
45              Temp_F = 1.8*Temp_C + 32;
46           run;

NOTE: The infile "/folders/myfolders/celsius.txt" is:
      Filename=/folders/myfolders/celsius.txt,
      Owner Name=root,Group Name=vboxsf,
      Access Permission=-rwxrwx---,
      Last Modified=16Dec2014:18:05:25,
      File Size (bytes)=10
```

The next section shows your program, along with information about your input data file. Notice that SAS Studio added an OPTIONS statement to your program. This statement controls what information is displayed in the log and how the data appears in the RESULTS window.

```
NOTE: 3 records were read from the infile "/folders/myfolders/celsius.txt".
      The minimum record length was 1.
      The maximum record length was 3.
NOTE: The data set WORK.CONVERT has 3 observations and 2 variables.
NOTE: DATA statement used (Total process time):
      real time           0.01 seconds
      cpu time            0.01 seconds
```

Here you see that three records (lines) were read from the input file.

```
NOTE: There were 3 observations read from the data set WORK.CONVERT.
NOTE: The PROCEDURE PRINT printed page 1.
NOTE: PROCEDURE PRINT used (Total process time):
      real time           0.03 seconds
      cpu time            0.04 seconds
```

Finally, you see that PROC PRINT read three observations. You also see the real and CPU times.

Before you close your SAS session, you should save your program. From the **CODE** tab, select the icon for **SAVE AS**:

Figure 7.10: The SAVE AS Icon

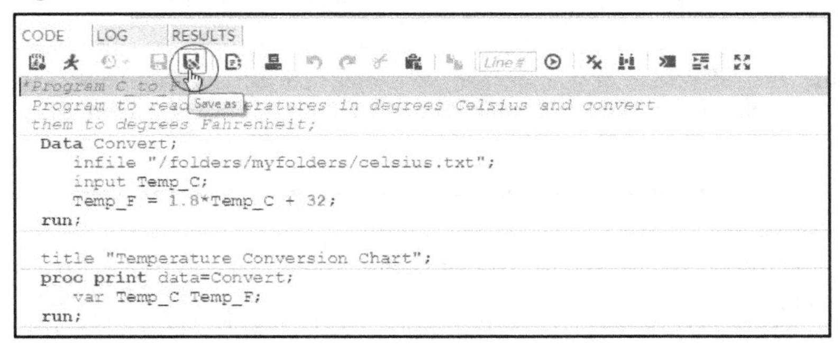

This brings up the following screen:

Figure 7.11: Saving Your Program in My Folders

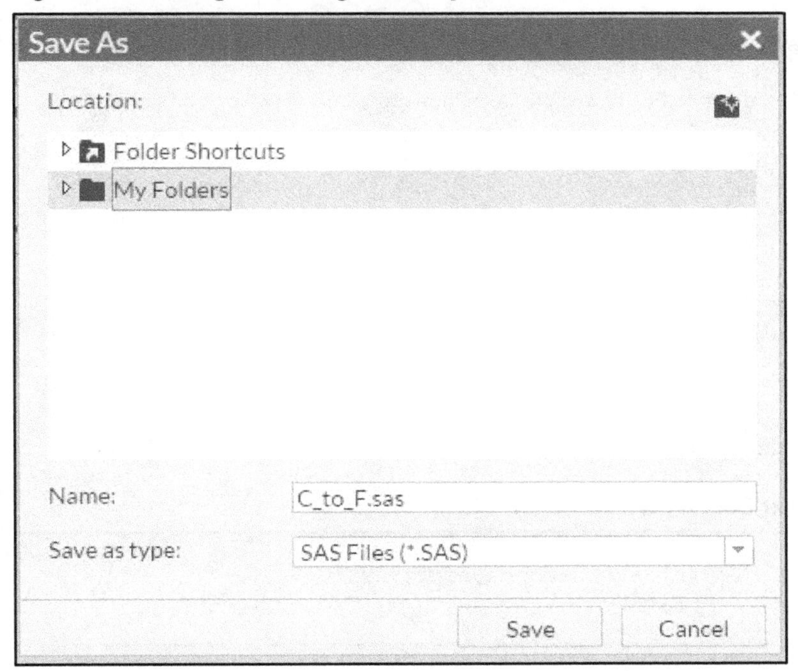

Enter the program name (in this example, it is C_to_F.sas) and click on the **Save** button. You can name your SAS program anything you like (as long as it meets the naming conventions for files on your computer). The extension .SAS is automatically added to the filename.

How the INPUT Statement Works

SAS programs can read just about any type of text data, whether it consists of numbers and letters, separated by delimiters (such as a CSV file), or whether the data values are arranged in fixed columns. This section just scratches the surface of the incredible versatility of this statement.

Reading Delimited Data

Let's start out by reading a file where data values (either character or numeric) are separated by delimiters. The following file, called Demographic.txt (stored in the `myfolders` folder), contains the following data values:

Variables in the File Demographic.txt

```
ID
Gender (M or F)
Age
Height (in inches)
Weight (in pounds)
Party (political party affiliation (I=Independent, R=Republican,
D=Democrat))
```

Here is the file:

File Demographic.txt

```
012345 F 45 65 155 I
131313 M 28 70 220 R
987654 F 35 68 180 R
555555 M 64 72 165 D
172727 F 29 62 102 I
```

You want to create a SAS data set called Demo from this raw data file. Proceed as follows:

Program 7.2: Reading Text Data from an External File

```
data Demo;
   infile '/folders/myfolders/Demographic.txt';
   input ID $ Gender $ Age Height Weight Party $;
run;
```

You simply list the variable names in the same order as the data values in the file. As you can probably guess, a dollar sign ($) following a variable name indicates that you want to read and store this value as character data. This is a good time to mention that SAS has only two variable types: character and numeric. By default, all numeric values are stored in 8 bytes (64 bits). In most programs, you specify the length of character variables. In this first program, because no character variable lengths are specified, a default length of 8 bytes (characters) is used to store each of the character values (even though Gender and Party are only one character in length).

Here is the SAS log that is produced when you run this program:

Figure 7.12: The SAS Log from Program 7.2

```
39         data Demo;
40            infile '/folders/myfolders/Demographic.txt/';
41            input ID $ Gender $ Age Height Weight Party $;
42         run;

NOTE: The infile '/folders/myfolders/Demographic.txt/' is:
      Filename=/folders/myfolders/Demographic.txt,
      Owner Name=root,Group Name=root,
      Access Permission=-rwxrwxrwx,
      Last Modified=26Dec2014:15:19:08,
      File Size (bytes)=108

NOTE: 5 records were read from the infile '/folders/myfolders/Demographic.txt/'.
      The minimum record length was 20.
      The maximum record length was 20.
NOTE: The data set WORK.DEMO has 5 observations and 6 variables.
NOTE: DATA statement used (Total process time):
      real time           0.01 seconds
      cpu time            0.02 seconds
```

You see information about the input data file, and note that 5 records were read (as expected) and the fact that a SAS data set, WORK.DEMO, was created. Unless you specify a storage location for your SAS data set, SAS places it in the WORK library. Data sets in the WORK library are temporary and disappear when you close your SAS session. You will see how to read and write permanent SAS data sets in Chapter 9. In the SAS log, you also see information on the real and CPU time used.

At this point, you can run PROC PRINT to see a listing of the file, or just click on the **LIBRARIES** tab, select the WORK library, and double-click the Demo data set. Here is what you will see:

Figure 7.13: Displaying the Data Set Using the SAS Studio Point-and-Click Feature

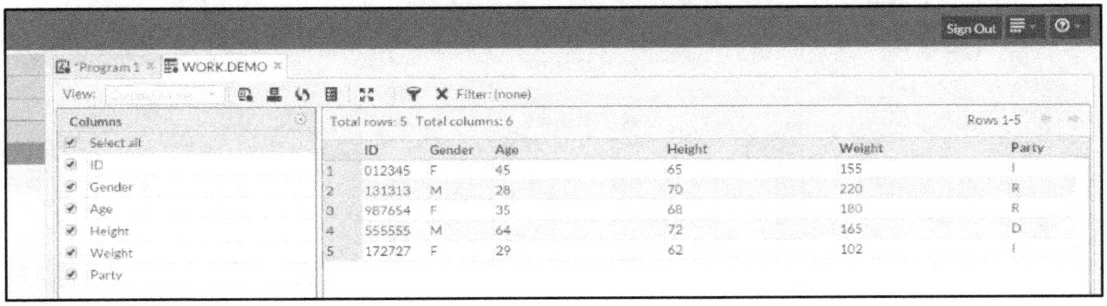

Sign Out

*Program 1 × WORK.DEMO ×

View: Filter: (none)

Columns: Select all, ID, Gender, Age, Height, Weight, Party

Total rows: 5 Total columns: 6 Rows 1-5

	ID	Gender	Age	Height	Weight	Party
1	012345	F	45	65	155	I
2	131313	M	28	70	220	R
3	987654	F	35	68	180	R
4	555555	M	64	72	165	D
5	172727	F	29	62	102	I

How Procedures (PROCS) Work

As you saw in the first part of this book, you can now use the **TASKS** tab to summarize the data or produce plots. However, because this is the programming section of the book, let's use PROC FREQ to compute frequencies and demonstrate how procedure options and statement options affect the results.

Program 7.3: Computing Frequencies on the Demo Data Set

```
data Demo;
   infile "/folders/myfolders/demographic.txt";
   input ID $ Gender $ Age Height Weight Party $;
run;

title "Computing Frequenciesfrom the DEMO Data Set";

proc freq data=Demo;
   tables Gender Party;
run;
```

You use PROC FREQ to compute frequencies on any of your variables. Use a TABLES statement to specify which variables you want to include in your results. These may include both character and numeric variables. If you include any numeric variables in the list, PROC FREQ will compute the frequency on every unique value—it does not group values into bins (there are other procedures that can produce histograms). If you do not use a TABLES statement, PROC FREQ will compute frequencies on all the variables in your data set.

The output from Program 7.3 is shown below:

Output from Program 7.3

Computing Frequenciesfrom the DEMO Data Set

The FREQ Procedure

Gender	Frequency	Percent	Cumulative Frequency	Cumulative Percent
F	3	60.00	3	60.00
M	2	40.00	5	100.00

Party	Frequency	Percent	Cumulative Frequency	Cumulative Percent
D	1	20.00	1	20.00
I	2	40.00	3	60.00
R	2	40.00	5	100.00

In the frequency tables you see frequencies, percent, cumulative frequencies, and cumulative percent for the two variables Gender and Party. The output you see results from the default settings for this procedure. In most cases, you will want to include procedure and statement options to control the output.

Most SAS procedures have what are called *procedure options*. These options affect how the procedure works, and they are placed between the procedure name and the semicolon. One popular procedure option used with PROC FREQ is ORDER=. There are several values you can select for this option. For this example, you will use ORDER=FREQ. This lists the frequencies from the most frequent to the least frequent in the output tables.

You will most likely use procedure statements with most procedures as well. The TABLES statement in Program 7.4 is an example of a procedure statement. You can also add statement options to control how a statement works. The rule is that statement options are placed after a slash (/) following the statement. The TABLES option NOCM (no cumulative statistics) is used in the next program to demonstrate a statement option. Let's see how the procedure option ORDER=FREQ and the statement option NOCUM affect the output:

Program 7.4: Adding Options to PROC FREQ

```
proc freq data=Demo order=freq;
   tables Gender Party / nocum;
run;
```

Your output now looks like this:

Output from Program 7.4

Adding Procedure and Statement Options

The FREQ Procedure

Gender	Frequency	Percent
F	3	60.00
M	2	40.00

Party	Frequency	Percent
I	2	40.00
R	2	40.00
D	1	20.00

The tables are now displayed in decreasing frequency order (notice the change in the order of Party), and cumulative statistics are no longer included in the tables. In most cases, you will want to include the NOCUM option on your TABLES statement.

How SAS Works: A Look Inside the "Black Box"

Although you can write basic SAS programs without understanding what goes on inside the "black box," a more complete understanding of how SAS works will make you a better programmer. Furthermore, this knowledge is essential when you are writing more advanced programs.

SAS processes the DATA step in two stages: In the first stage, called the *compile stage*, several activities take place. The SAS complier reads each line of code from left to right, top to bottom. Each statement is broken up into tokens, with certain keywords such as DATA and RUN causing certain actions to take place. What is important to you as a programmer is to know that this is the stage where the data descriptor for each of your variables is written out. The first time SAS encounters a variable in a DATA step, it decides if that variable is numeric or character. If it is numeric, SAS gives it a default length of 8 bytes. If it is character, it has rules that it uses to determine the storage length. SAS character values can be a maximum of 32,767 bytes in length. In this first stage, your source code is also compiled into machine language. Also during the compile stage, SAS determines which variables will be written out to the new data set and which variables will be dropped (i.e., not written out). One way of determining whether a variable is kept or dropped is by explicitly writing a KEEP or DROP statement in the DATA step or by including a KEEP= or DROP= data set option (more on that later).

In the second stage, called the *execute stage*, the program performs its functions of reading data, performing logical actions, iterating loops, and so on. When you are reading raw data from a file or if you have variables defined in assignment statements (such as the variable Temp_F in Program 7.1), SAS initializes each of these variables with a missing value at the top of the DATA step. During execution of the DATA step, these variables are usually given values, either from the raw data or from a computation. At the bottom of the DATA step (defined by the RUN statement), SAS performs an automatic output to one or more data sets. The DATA step continues its internal loop, reading data, performing calculations, and outputting observations to a data set. If you are reading data from a text file or from a SAS data set, the DATA step will stop when you reach the end of the file on any file.

Conclusion

At this point, you understand how to write a simple program using the SAS University Edition. You understand the various windows inside SAS Studio. You can read external data where blanks are used as delimiters and produce simple reports. The next chapter explores the INPUT statement, one of the most powerful statements in SAS.

Chapter 8: Reading Data from External Files

Introduction

This chapter describes three of the most common methods that you can use to read raw text data using SAS. It also demonstrates how to create SAS data sets from Excel workbooks. If you typically receive data that is already in SAS data sets, you can skip this chapter (unless you are just curious).

Reading Data Values Separated by Delimiters

One common method of storing data in text files is to separate data values by a delimiter, usually blanks or commas. SAS refers to this as *list input*.

Let's start out with a file where blanks (spaces) are used as delimiters. This is a good starting place because a blank is the default delimiter in SAS. In this example, the data file you want to read contains an ID, first name, last name, gender, age, and state abbreviation. These data lines are stored in a file called Blank_Delimiter.txt stored in the `c:\SASUniversityEdition\myfolders` directory on your hard drive. Here is the listing:

File Blank_Delimiter.txt (located in c:\SASUniversityEdition\myfolders)

```
103-34-7654 Daniel Boone M 56 PA
676-10-1020 Fred Flintstone M
454-30-9999 Tracie Wortenberg F 34 NC
102-87-8374 Jason Kid M 23 NJ
888-21-1234 Patrice Marcella F . TX
788-39-1222 Margaret Mead F 77 PA
```

A program to read this data file is shown next:

Program 8.1: Reading Data with Delimiters (Blanks)

```
/* This is another way to insert a comment */

data Blanks /* the data set name is Blanks */;
   infile "/folders/myfolders/Blank_Delimiter.txt" missover;
   informat ID $11. First Last $15. Gender $1. State_Code $2.;
   input ID First Last Gender Age State_Code;
run;

title "Listing of Data Set BLANKS";
proc print data=blanks;
run;
```

The first line of this program demonstrates another way to insert a comment in a SAS program. You begin the comment with a /* and end the comment with a */. Unlike a comment statement that begins with an asterisk, this type of comment can be placed within a SAS statement (as demonstrated in the DATA statement of this program.).

The INFILE statement tells the SAS program where to find the input data. In this example, because the data file was placed in the location predefined by SAS Studio as myfolders, you specify the file location as shown (even though the actual file on your computer is **`c:\SASUniversityEdition\myfolders`**.

Notice the keyword MISSOVER on this statement. This is not necessary if you have data values for every variable for every line of data. Line 2 of the data file (Fred Flintstone) is missing the last two values (Age and State_Code). Without the MISSOVER option, SAS would try to read these two values from the next line of data! Obviously, you never want this to happen. The MISSOVER option comes into play when there are missing values at the end of the data line and you are using list input. This option tells the program to set each of these variables to a missing value. Notice line 5 (Patrice Marcella). To indicate that there is a missing value for Age, you use a period. Without the period, SAS would try to read the next value (TX) as the Age and really screw things up. This is handled differently in the next section describing CSV (comma-separated values) files. One last point: SAS interprets multiple blanks as a single delimiter.

You list each of the variable names on the INPUT statement in the order in which they appear in the raw data file. When you use list input (the type of input that reads delimited data), the default length for character variables is 8. To specify how to read your character variables, you use an INFORMAT statement. Following the keyword INFORMAT, you list one or more variables (usually character variables) and follow the variable or variables with a dollar sign (that indicates a character variable) and the number of characters (bytes) of storage you want, followed by a period. In this example, $11., $15., $1., and $2. are called *character informats*. Informats are used in other styles of input as well.

Use of these informats also determines the storage length of each of the variables listed in the INFORMAT statement. In this example, you are specifying a length of 11 for ID, a length of 15 for both the variables First and Last, a length of 1 for Gender, and a length of 2 for State_Code. You could also define the lengths of these character variables with a LENGTH statement. That statement would be similar to the INFORMAT statement—just replace the keyword INFORMAT with the keyword LENGTH and leave off the periods after the digits. An INFORMAT statement can also be used to tell SAS how to read other types of data such as dates and numbers with commas and dollar signs, therefore making it more flexible than a LENGTH statement for instructing SAS on how to read raw data.

To list the observations in the Blanks data set, you use PROC PRINT. Below are sections from the SAS log:

Figure 8.1: SAS Log from Program 8.1

```
58         data blanks;
59            infile "/folders/myfolders/Blank_Delimiter.txt" missover;
60            informat ID $11. First Last $15. Gender $1. State_Code $2.;
61            input ID First Last Gender Age State_Code;
62         run;

NOTE: The infile "/folders/myfolders/Blank_Delimiter.txt" is:
      Filename=/folders/myfolders/Blank_Delimiter.txt,
      Owner Name=root,Group Name=root,
      Access Permission=-rwxrwxrwx,
      Last Modified=30Dec2014:11:41:06,
      File Size (bytes)=205

NOTE: 6 records were read from the infile "/folders/myfolders/Blank_Delimiter.txt".
      The minimum record length was 29.
      The maximum record length was 37.
```

```
NOTE: The data set WORK.BLANKS has 6 observations and 6 variables.
NOTE: DATA statement used (Total process time):
      real time           0.01 seconds
      cpu time            0.02 seconds
```

You see that there were no errors in the program, along with information on your input file. Finally, you see that your data set (Blanks) was created with 6 observations and 6 variables, along with the real and CPU time.

Even when you see the OUTPUT window after you submit a program, it is a good idea to check the SAS Log for messages and warnings.

Here is the output:

Output from Program 8.1

Listing of Data Set BLANKS

Obs	ID	First	Last	Gender	State_Code	Age
1	103-34-7654	Daniel	Boone	M	PA	56
2	676-10-1020	Fred	Flintstone	M		.
3	454-30-9999	Tracie	Wortenberg	F	NC	34
4	102-87-8374	Jason	Kid	M	NJ	23
5	888-21-1234	Patrice	Marcella	F	TX	.
6	788-39-1222	Margaret	Mead	F	PA	77

Everything looks fine. Notice the missing values in observations 2 and 5. In observation 2, the missing values for State_Code and Age result from the MISSOVER option on the INFILE statement. The missing value for Age in observation 5 results from the period in the input data.

Reading Comma-Separated Values Files

CSV files (comma-separated values) are one of the most common file types for delimited data. One common use of CSV files is to output data from an Excel workbook. CSV files use commas to separate data values and, unlike the default behavior of SAS when you use blank delimiters, CSV files interpret two commas in a row to mean that there is a missing value for that data field.

The CSV file in this example contains the same data as the file Blank_Delimiter.txt used in the last example. Here is a listing of the file:

File Comma_Delimiter.txt

```
103-34-7654,Daniel,Boone,M,56,PA
676-10-1020,Fred,Flintstone,M
454-30-9999,Tracie,Wortenberg,F,34,NC
102-87-8374,Jason,Kid,M,23,NJ
888-21-1234,Patrice,Marcella,F,,TX
788-39-1222,Margaret,Mead,F,77,PA
```

Notice the two commas in a row for the fifth line of data (Patrice Marcella). Unlike the blank-delimited file where you needed a period to indicate a missing value for Age, the two commas in a row indicate that this value is missing.

To read data from a CSV file, you add the DSD (Delimiter-Sensitive Data) option on the INFILE statement. This option does several things: First, it understands that the data values in the file are separated by commas. Next, it understands that two commas indicate a missing value. Also, if you have a data value in quotation marks (for example, a state name like 'New York' that consists of two words separated by a blank space), it will ignore any delimiters inside the quotes and strip the quotes when it assigns the value to a SAS variable. The program to read the data from the CSV file Comma_Delimiter.txt is shown next:

Program 8.2: Reading CSV Files

```
data commas;
   infile "/folders/myfolders/Comma_Delimiter.txt" dsd missover;
   informat ID $11. First Last $15. Gender $1. State_Code $2.;
   input ID First Last Gender Age State_Code;
run;

title "Listing of Data Set COMMAS";
proc print data=commas;
run;
```

Output from this program is identical to the output from Program 8.1.

Reading Data Separated by Other Delimiters

By using the DLM= option on the INFILE statement, you can specify any delimiter you wish, including non-printing characters such as tabs. Suppose you have a tab-delimited file that you want to read. Because a tab character is a non-printing character, you need to find the hexadecimal code for a tab in ASCII (the coding system used in Windows, UNIX, and Linux systems). In ASCII, the Hex representation for a tab is 09. You can specify a Hex character anywhere in a SAS program by using a Hex constant in the form '*nn*'x, where *nn* is the Hex value you want (placed in single or double quotes) and the 'x' is in lowercase or uppercase. Also, there are no spaces between the quoted Hex value and the 'x'. The program below reads an ASCII file where a tab was used as the delimiter.

Program 8.3: Reading Tab-Delimited Data

```
data tabs;
   infile "/folders/myfolders/Tab_Delimiter.txt" dlm='09'x missover;
   informat ID $11. First Last $15. Gender $1. State_Code $2.;
   input ID First Last Gender Age State_Code;
run;

title "Listing of Data Set TABS";
proc print data=tabs;
run;
```

The output is identical to the output from the previous two programs. You can use the DLM= option on the INFILE statement to specify any delimiter you wish. If you want two consecutive delimiters to indicate a missing value, include the DSD option as well as the DLM= option. For example, if you have data that uses a pipe symbol (vertical bar) as a delimiter and you want to interpret two bars together to indicate that there is a missing value, your INFILE statement would look like this:

```
infile "file-location" dlm='|' dsd;
```

Reading Data in Fixed Columns

Another very common method of placing data in text files is to assign values to predefined columns. SAS provides you with two methods of reading column data. One is called *column input*. With this method, you follow a variable name with a starting and ending column. If the variable only occupies one column, you do not specify an ending column. If the variable will be defined as a character variable, you place a dollar sign ($) between the variable name and the starting column number. This method is restricted to reading standard numeric data (numbers with or without decimal points) and character data.

The other method is called *formatted input*. With this method, you specify a starting column, the variable name, and what SAS calls an *informat*. SAS informats give SAS information on how to read data from one or more columns. Formatted input is much more flexible than column input because it can read data values such as dates and times. You will see examples of both of these methods in the sections that follow.

Column Input

For this example, you want to read data from a file called Health.txt. This file contains the following variables:

Variables in the Health.txt Data File

Variable Name	Description	Starting Column	Ending Column
Subj	Subject number	1	3
Gender	Gender (M or F)	4	4
Age	Age in years	5	6
HR	Heart rate	7	8
SBP	Systolic blood pressure	9	11
DBP	Diastolic blood pressure	12	14
Chol	Total cholesterol	16	18

The data file looks like this:

File Health.txt (located in c:\SASUniversityEdition\myfolders)

```
123456789012345678901234567890 (Ruler - this line is not in the file)
001M2368120 90128
002F5572180 90170
003F1858118 72122
004M8082      220
005F3462128 80
006F3878108 68220
```

Because the data values are in fixed columns, you can use column input to read it. Notice that blanks are used when there are missing values (although there are no blanks at the end of short lines). Below is a program that reads this data file using column input.

Program 8.4: Reading Data in Fixed Columns Using Column Input

```
data Health;
   infile '/folders/myfolders/health.txt' pad;
   input Subj   $ 1-3
         Gender $ 4
         Age      5-6
         HR       7-8
         SBP      9-11
         DBP      12-14
         Chol     15-17;
run;
```

```
title "Listing of Data Set HEALTH";
proc print data=Health;
   ID Subj;
run;
```

The INFILE statement tells the program where to find the Health.txt data file. Following the file location, you see the keyword PAD. **This is very important**. Because some of the lines in the file are shorter than others (there is a carriage return after the last number in each line), SAS will not read the data correctly without it. This option pads each line with blanks. In the "old" days when all programming was done on mainframe computers (often using punch cards as input), data lines were automatically padded with blanks (typically up to 80 columns). Because a lot of data is entered on personal computers, data lines can be much longer than 80 columns and they are not typically padded with blanks.

Here is a listing of data set HEALTH:

Output from Program 8.4

Listing of Data Set HEALTH

Subj	Gender	Age	HR	SBP	DBP	Chol
001	M	23	68	120	90	128
002	F	55	72	180	90	170
003	F	18	58	118	72	122
004	M	80	82	.	.	220
005	F	34	62	128	80	.
006	F	38	78	108	68	220

Notice that the blanks in the raw data file result in missing values in the listing.

Formatted Input

Formatted input is the most common (and flexible) method for reading data in fixed columns. Let's jump right to the program and then the explanation. Here it is:

Program 8.5: Reading Data in Fixed Columns Using Formatted Input

```
data health;
   infile '/folders/myfolders/health.txt' pad;
   input @1  Subj   $3.
         @4  Gender $1.
         @5  Age     2.
         @7  HR      2.
         @9  SBP     3.
         @12 DBP     3.
```

```
          @15 Chol    3.;
run;

title "Listing of Data Set HEALTH";
proc print data=health;
   ID Subj;
run;
```

The @ sign is called a *column pointer*. The number following the @ sign is the starting column for the value you want to read. Following the variable name is an informat. There are lots of informats for reading and interpreting things like dates, times, and values with dollar signs and commas. This program only uses two: The informat $*w*. (*w* stands for width) reads *w* columns of character data; the informat *w*. reads *w* columns of numeric data. The informat for numeric data is actually more general. You can specify a numeric informat of *w.d*, where *w* is the total number of columns to read and *d* indicates that there is a decimal place with *d* digits to the right. When you use a *w.d* format, if the value you are reading contains a decimal point, the *d* in the informat is ignored. The data set created by this program is identical to the data set produced by Program 8.4.

There are some shortcuts you can use when employing formatted input. One of the most useful shortcuts utilizes variable lists and informat lists. If you have a group of variables that all share the same informat, you can list all the variables together (in a set of parentheses) and follow the list of variables by one or more informats (also in a set of parentheses). Here is Program 8.5, rewritten using this feature.

Program 8.6: Demonstrating Formatted Input

```
data health;
   infile '/folders/myfolders/health.txt' pad;
   input @1  Subj   $3.
         @4  Gender $1.
         @5  (Age HR) (2.)
         @9  (SBP DBP Chol) (3.);
run;

title "Listing of Data Set HEALTH";
proc print data=health;
   ID Subj;
run;
```

In this program, Age and HR both use the 2. informat; SBP, DBP, and Chol all use the 3. informat.

> Technical note: If you have as many informats as you have variables, the variables and informats will "pair up" on a one-to-one basis. If there are fewer informats than variables, SAS will go back to the beginning of the informat list and reuse the informats in order. If there is only one informat, it will apply to every variable in the variable list. This is by far the most common way variable lists and informat lists are used.

In this example, this shortcut only saved a few lines of typing. However, imagine that you had 50 character values (called Ques1-Ques50), each occupying one column. You could write a very compact INPUT statement like this:

```
input (Ques1-Ques50($1.);
```

The variable list also demonstrates a convenient way to reference all the variables from Ques1 to Ques50. Anywhere in a SAS program where you need to name variables that have the same alphabetic root (Ques in this example), you can use a single dash to indicate that you are referencing all the variables from the first to the last.

Reading Excel Files

One of the most common sources of data is Excel workbooks (either XLS or XLSX files). SAS Studio has some easy ways to read this type of data and create SAS data sets. One of the easiest ways to read Excel data is with the built-in Utility option on the **Tasks** tab of SAS Studio. This section shows how to read Excel data using SAS code and a built-in snippet.

To demonstrate how this works, an Excel workbook contains student information (name and ID) along with some quiz and exam grades. It looks like this:

Figure 8.2: Excel Workbook Containing Student Information and Grades

	A	B	C	D	E	F	G	H
1	Name	ID	Quiz1	Quiz2	Midterm	Quiz3	Quiz4	Final
2	Jones	12345	88	80	76	88	90	82
3	Hildebran	22222	95	92	91	94	90	96
4	O'Brien	33333	76	78	79	81	83	80
5								
28								
29								
30								

Sheet1

The first row of the table contains column headings. These headings will become variable names in the SAS data set. For this example, all of the column headings are valid SAS variable names. You will see later what happens when this is not the case. This worksheet was saved with the default name Sheet1.

To read this workbook and create a SAS data set, click on the **Snippets** tab in SAS Studio and select **Data** from the drop-down list: It should look like this:

Figure 8.3: The Data Tab under Snippets

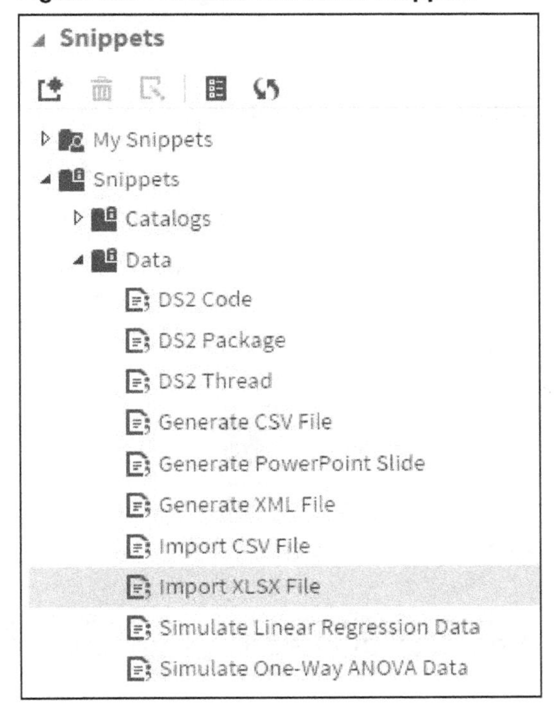

Select **IMPORT XLSX FILE**. (Note: Use this for either XLS or XLSX files.) This brings up the following code snippet:

```
PROC IMPORT DATAFILE="<Your XLS File>"
                    OUT=WORK.MYEXCEL
                    DBMS=XLSX
                    REPLACE;
RUN;

/** Print the results. **/

PROC PRINT DATA=WORK.MYEXCEL; RUN;
```

All you need to do is enter the name of your Excel workbook. If the file is an older XLS file instead of the newer XLSX file, replace the **DBMS=XLSX** with **DBMS=XLS**. The grades workbook was saved in the location **SASUniversityEdition\myfolders**, so your import program should look like this:

Program 8.7: File Snippet to Import the Grades.xlsx Workbook

```
PROC IMPORT DATAFILE="/folders/myfolders/grades.xlsx"
                   OUT=WORK.GRADES
                   DBMS=XLSX
                   REPLACE;
RUN;

/** Print the results. **/

PROC PRINT DATA=WORK.Grades; RUN;
```

Submit this program and you have a temporary SAS data set called Grades. Results of the PROC PRINT look like this:

Listing of data set Grades

Obs	Name	ID	Quiz1	Quiz2	Midterm	Quiz3	Quiz4	Final
1	Jones	12345	88	80	76	88	90	82
2	Hildebrand	22222	95	92	91	94	90	96
3	O'Brien	33333	76	78	79	81	83	80

As an alternative, you can save your workbook as a CSV file and use the **INPUT CSV** tab instead of the **INPUT XLSX** tab.

Reading from an Excel Workbook where Column Headings Are Invalid SAS Variable Names

It is quite possible, indeed probable, that the person giving you data in an Excel workbook does not know the rules for naming SAS variables. To see what happens when you import a workbook with invalid SAS variable names as column headings, take a look at the following workbook:

Figure 8.4: Excel Workbook Grades2.xlsx

	A	B	C	D	E	F	G	H
1	Stuent Name	ID	Quiz 1	Quiz 2	Mid Term	Quiz 3	Quiz 4	2015Final
2	Jones	12345	88	80	76	88	90	82
3	Hildebrand	22222	95	92	91	94	90	96
4	O'Brien	33333	76	78	79	81	83	80

With the exception of ID, all of the column headings are invalid SAS variable names. Many of the names contain spaces. 2015Final is invalid because it begins with a digit. Let's use the import snippet to convert this worksheet into a SAS data set and see what happens.

You click on the **Snippets** tab, select **IMPORT XLSX**, and edit the program so that it looks like this:

Program 8.8: Code to Import the Grades2 Workbook

```
/** Import an XLS file.   **/

PROC IMPORT DATAFILE="/folders/myfolders/Grades2.xlsx"
                  OUT=WORK.Grades2
                  DBMS=XLSX
                  REPLACE;
RUN;

/** Print the results. **/

PROC PRINT DATA=WORK.Grades2; RUN;
```

Here is the listing:

Output from Program 8.8

Obs	Stuent_Name	ID	Quiz_1	Quiz_2	Mid_Term	Quiz_3	Quiz_4	_2015Final
1	Jones	12345	88	80	76	88	90	82
2	Hildebrand	22222	95	92	91	94	90	96
3	O'Brien	33333	76	78	79	81	83	80

The blanks were replaced by an underscore and an underscore was added as the first character in the column heading 2015Final to create valid SAS variable names.

Conclusion

This chapter explored how to read external text data in almost any format. This is only the tip of the iceberg. For more information on the INPUT statement, please check out *Learning SAS by Example* (Cody, 2007), published by SAS Press.

Problems

You can download all the files and programs you need for these problems from the author's web site: support.sas.com/cody.

1. A quick survey was conducted, and the following data values were collected:

Variable	Description
Subj	Subject number (3 digits – stored as character)
Gender	F=Female, M=Male
DOB	Date of birth in *mm/dd/yyyy* form
Height	Height in inches
Weight	Weight in pounds
Income_Group	Income group: L=Low, M=Medium, H=High

The data values were saved in a blank-delimited file called Quick.txt. Copy this file to a folder called `c:\SASUniversityEdition\myfolders` and write a SAS program to read this data file, create a temporary SAS data set, and produce a listing of the file. The file appears as follows:

Figure 5: Listing of File Quick.txt

```
001 M 10/21/1950 68 150 H
002 F 9/11/1981 63 101 M
003 F 1/1/1983 62 120 L
004 M 5/17/2000 57 98 L
005 M 7/15/1970 79 220 H
006 F 6/1/1968 71 188 M
```

Use the mmddyy10. informat to read the DOB. Also, include the following statement in your program:

```
format DOB mmddyy10.;
```

The reason for this is that, as you will see later in the chapter on dates, SAS stores dates as the number of days from January 1, 1960. The statement above is an instruction to print the DOB as a date and not the internal value (the number of days from 1/1/1960).

2. Use PROC FREQ to compute frequencies for the variables Gender and Income_Group.

3. Add options to your program in Problem 2 to have the table list values in decreasing order of frequency and omit the cumulative statistics from the tables.
4. Modify the program in Problem 1 to compute a new variable called BMI (body mass index). BMI is defined as the weight in kilograms divided by the height (in meters) squared. Conversions are:

   1 kg. = 2.2 pounds

   1 meter = 39.3701 inches
5. The same data described in Problem 1 was saved as a CSV file called Quick.csv. Write a program to create a SAS data set from this CSV file. Be sure to include the FORMAT statement mentioned in that problem.
6. The same data described in Problem 1 in Chapter 8 was entered with the forward slash (/) as a delimiter. Because the dates also include slashes, the dates were placed in quotes. Here is a listing of the file:

   Listing of File Quick_Slash.txt

   ```
   001/M/"10/21/1950"/68/150/H
   002/F/"/9/11/1981"/63/101/M
   003/F/"1/1/1983"/62/120/L
   004/M/"5/17/2000"/57/98/L
   005/M/"7/15/1970"/79/220/H
   006/F/"6/1/1968"/71/188/M
   ```

   The variables in the file are ID (allow for 11 characters), First and Last name (each up to 15 characters), Gender (M or F), Age, and State code (2 characters). Write a program to read data from this file, create a temporary SAS data set (call it Slash), and produce a listing of the file.

   Hints: 1) Some lines are missing values at the end of the line, and 2) Two slashes in a row indicate there is a missing value (think DSD).
7. An Excel workbook called Grades.xlsx contains data on student grades. Use the Import XLS snippet (in the **Snippet** task on the navigation pane) to read the spreadsheet and create a SAS data set.
8. The data from the quick survey was entered into a file called Quick_Cols.txt using fixed columns as follows:

Variable	Description	Columns
Subj	Subject number (3 digits)	1-3
Gender	F=Female, M=Male	4
DOB	Date of birth in *mm/dd/yyyy* form	5-14

Height	Height in inches	15-16
Weight	Weight in pounds	17-19
Income_Group	Income group: L=Low, M=Medium, H=High	20

Using column input, create a SAS data set from this file. Important note: Because DOB is a date, you will have to read it as a character string.

9. Create a SAS data set from the file Quick_Cols.txt using formatted input. Read the DOB with the mmddyy10. informat. Also, include the FORMAT statement mentioned in Problem 1.
10. What's wrong with this program?

```
1. data Names;
2.   input Name $ Height Weight;
3.   Height_CM = Heightx2.54;
4.   *Note 1 inch = 2.54 cm
5. datalines;
   Zemlachenko 73 190
   Holland 63 100
   ;
```

Chapter 9: Reading and Writing SAS Data Sets

What's a SAS Data Set?

Besides reading raw data from text files, SAS can also read and write data from SAS data sets. Once you have created a SAS data set from raw data or you have been given a SAS data set, you are ready to run SAS procedures to analyze your data or to write a DATA step to create new variables or to further manipulate your data.

A SAS data set actually contains two parts: One is called the data descriptor; the other is the data itself. The data descriptor contains your variable names, whether a variable is stored as character or numeric (the only two types allowed in a SAS data set), how many bytes of storage are used to store a variable, and other information about how to display the variable in reports or charts. A fancy word used to describe the data descriptor portion of a SAS data set is *metadata*—data about your data.

The format of a SAS data set is proprietary to SAS and only SAS can read and write SAS data sets directly (there are other programs that can read SAS data sets and convert them to other formats). If you try to open a SAS data set in Word or Notepad, it will look like gibberish.

If you want to examine the data descriptor (metadata) for a SAS data set, you have several options. Since this is the programming portion of the book, the method described here is to run a SAS procedure called PROC CONTENTS. Suppose you want to see the data descriptor for the data set called Demo described in Chapter 7.

The program shown next uses PROC CONTENTS to do this:

Program 9.1: Running PROC CONTENTS to Examine the Data Descriptor for Data Set Demo

```
*Program to display the data descriptor of data set DEMO;

title "Data Descriptor for Data Set DEMO";
proc contents data=demo;
run;
```

You use a DATA= procedure option to specify which data set you want to examine. Here is the output from this procedure:

Output from Program 9.1

Data Descriptor for Data Set DEMO

The CONTENTS Procedure

Data Set Name	WORK.DEMO	Observations	5
Member Type	DATA	Variables	6
Engine	V9	Indexes	0
Created	01/10/2015 10:49:35	Observation Length	48
Last Modified	01/10/2015 10:49:35	Deleted Observations	0
Protection		Compressed	NO
Data Set Type		Sorted	NO
Label			
Data Representation	SOLARIS_X86_64, LINUX_X86_64, ALPHA_TRU64, LINUX_IA64		
Encoding	utf-8 Unicode (UTF-8)		

The first section of output shows global information about your data set: the number of observations, the number of variables, etc. You also see the date and time the data set was created or modified.

Engine/Host Dependent Information	
Data Set Page Size	65536
Number of Data Set Pages	1
First Data Page	1
Max Obs per Page	1360
Obs in First Data Page	5
Number of Data Set Repairs	0
Filename	/tmp/SAS_work3B6B0000505C_localhost.localdomain/SAS_workF5E70000505C_localhost.localdomain/demo.sas7bdat
Release Created	9.0401M2
Host Created	Linux
Inode Number	144292
Access Permission	rw-rw-r--
Owner Name	sasdemo
File Size (bytes)	131072

The next section of your output contains technical information about the size of the data set, the page size, the access permissions, and other details that you may (or may not) be interested in.

Alphabetic List of Variables and Attributes			
#	Variable	Type	Len
3	Age	Num	8
2	Gender	Char	8
4	Height	Num	8
1	ID	Char	8
6	Party	Char	8
5	Weight	Num	8

This last section of the output lists each of your variable names (in alphabetical order), the file type (Num for numeric, Char for character), and the number of bytes of storage used.

Temporary versus Permanent SAS Data Sets

All of the data sets described in the previous chapter were temporary data sets—that is, once you end your SAS session, these data sets disappear. To create or read a permanent SAS data set, you need to specify two parts: the location where the data set is stored (referred to as a *library* in SAS terminology) and the actual data set name. You can specify a permanent SAS data set name by specifying the library name, a period, and then the data set name. Permanent SAS data sets are in the form:

```
library-reference.data-set-name
```

where *library-reference* (referred to as a *libref* in SAS terminology) is an alias for a folder and *data-set-name* is the name of the SAS data set. You may wonder about the data sets you created in the last chapter. Those data set names did not contain a period. It turns out that when you use a single name on a DATA statement (such as `DATA Test;`), SAS assumes that the data set is a temporary data set and stores it in the Work library (a temporary folder/directory located in your user area). The true name of the data set is `Work.Test`, where `Work` is the built-in library reference to the Work library.

There are two steps in creating or reading a permanent SAS data set. Step one is to create the libref by using a LIBNAME statement or by using the SAS Studio point-and-click interface. If you are programming in a non-virtual environment, this process is straightforward. For example, if you want to create a permanent SAS data set in the `c:\mydata folder` on your computer, you first create a library reference (the name Oscar is used in this example) like this:

```
libname Oscar 'c:\myfolder';
```

Library references are a maximum of 8 characters and follow SAS naming conventions. If you want to create a permanent SAS data set called Demo in the `c:\myfolder folder`, your DATA statement looks like this:

```
data Oscar.Demo;
```

Data set Demo will remain even after you close your SAS session. If you use your operating system to list all the files in the `c:\mydata folder`, you will see a file called Demo.sas7bdat (the extension refers to a SAS data set that is a binary data file compatible with SAS 7).

Unfortunately, referencing a file on your hard drive from within your virtual machine makes things a bit more difficult, hence the next section.

Shared Folders: Communicating between Your Virtual Computer and Your Real Computer

Here's the problem. SAS University Edition runs SAS on a virtual machine. The virtual machine uses the Linux operating system. Filenames on Linux (or UNIX for that matter) use a forward slash (/) to specify folders and subfolders and the filenames are case-sensitive. Computers running a Microsoft operating system use a backslash (\) to specify folders and subfolders. Also, filenames on PCs or Macs are not case-sensitive. The way that SAS University Edition decided to solve this incompatibility was to use shared folders. You first create shared folders on your virtual machine. Then, inside SAS Studio, these shared folders show up as folder shortcuts for you to use.

The good news is that if you followed the SAS University Edition setup instructions, there is a default shared folder called `myfolders` that points to a specific folder on your hard drive. The suggestion is a folder called `\SASUniversityEdition\myfolders`. So, an example on Windows could be:

```
c:\SASUniversityEdition\myfolders
```

In Program 8.1 in the previous chapter, you had a text file called Celsius.txt located in `c:\SASUniversityEdition\myfolders`. The INFILE statement to read that file was:

```
infile '/folders/myfolders/celsius.txt';
```

If you want to read or write SAS data sets to the `myfolders` folder, you write a LIBNAME statement such as:

```
libname Oscar '/folders/myfolders';
```

As an alternative, you can right-click on a folder where you wish to store your SAS data sets. It looks like this:

Figure 9.1: Using SAS Studio to Create a Library

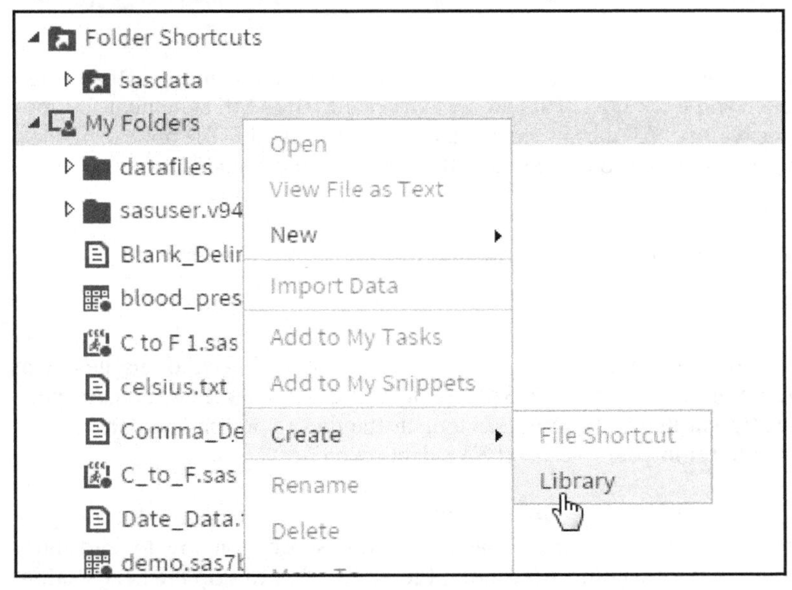

Move the pointer to **Create** and select **Library**. This brings up the following dialog box:

Figure 9.2: The Next Step in Creating a SAS Library Using SAS Studio

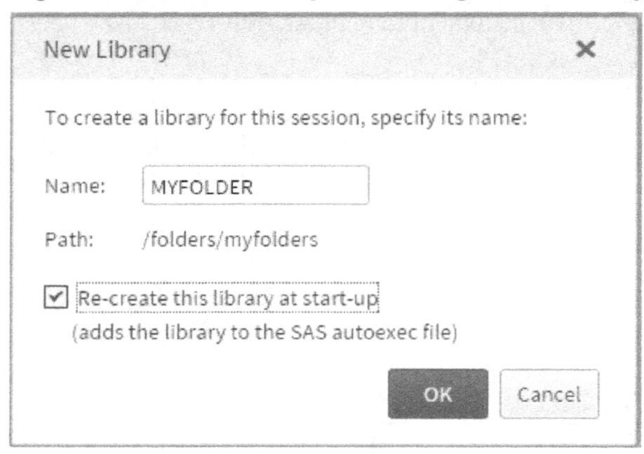

You can leave the library name as is or choose another name. Then be sure the box next to **Re-create this library at start-up** is checked and click **OK**. This process writes a LIBNAME statement for you and stores it in a file called Autoexec.sas. All the statements that are located in this file are executed automatically every time you open a SAS session. As you will see later, you can add other statements such as options to this file.

To create a permanent SAS data set called Demo, you write the following DATA statement:

```
data Oscar.Demo;
```

If you have not saved the LIBNAME statement in the Autoexec.sas file, you will need to reissue any of your LIBNAME statements every time you open a new SAS session. It doesn't matter what name you choose for your library reference (Oscar in this example), as long as the libref you use on your LIBNAME statement matches the libref in your two-level SAS data set name.

It may be inconvenient to move all of your data files to the `SASUniversity\myfolders` location, especially if you have many files or they are very large. The next section shows you how to read and write to any folder on your hard drive from within SAS Studio. Please feel free to skip the next section if you plan to always use the `\SASUniversity\myfolders` location for your SAS University Edition data.

Creating a Shared Folder and Reading Data from It

The best way to demonstrate creating a shared folder is to show you screen shots of each stage in the process. The screen shots were made using Oracle VirtualBox. The process in VMWare Workstation Player is similar, the only difference being the commands to create the shared folders.

In this example, you have a file called HeightWeight.txt located on your hard drive in a folder called `c:\sasdata`.

You start by creating a shared folder on your virtual machine.

Click on the **Settings** icon.

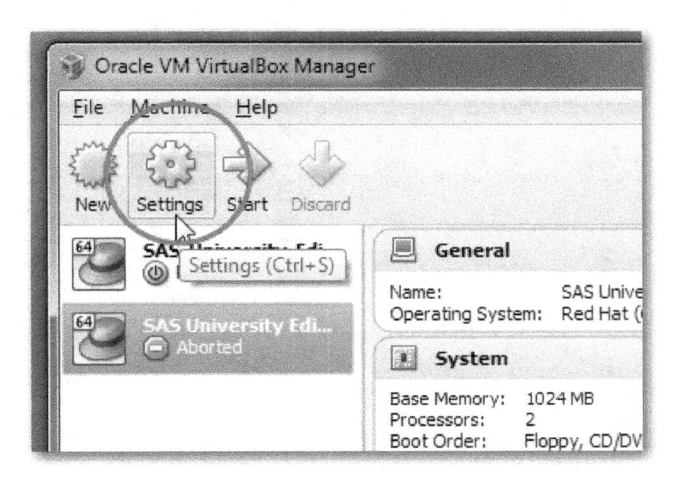

Under the Settings menu, select **Shared Folders**.

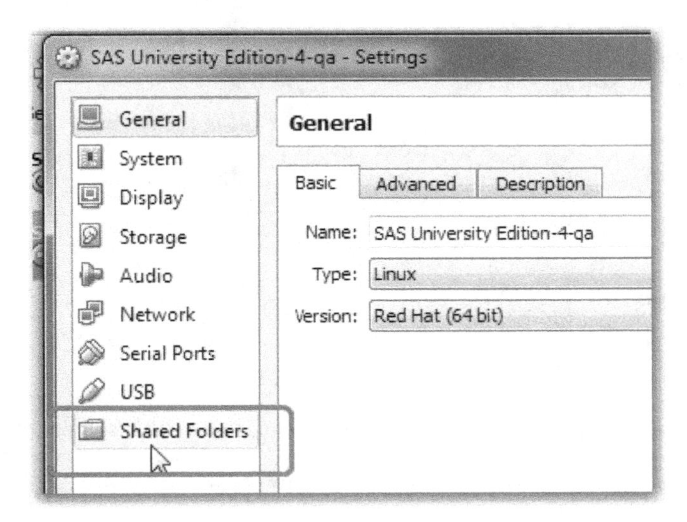

To add a new folder, click the plus (+) sign as shown below:

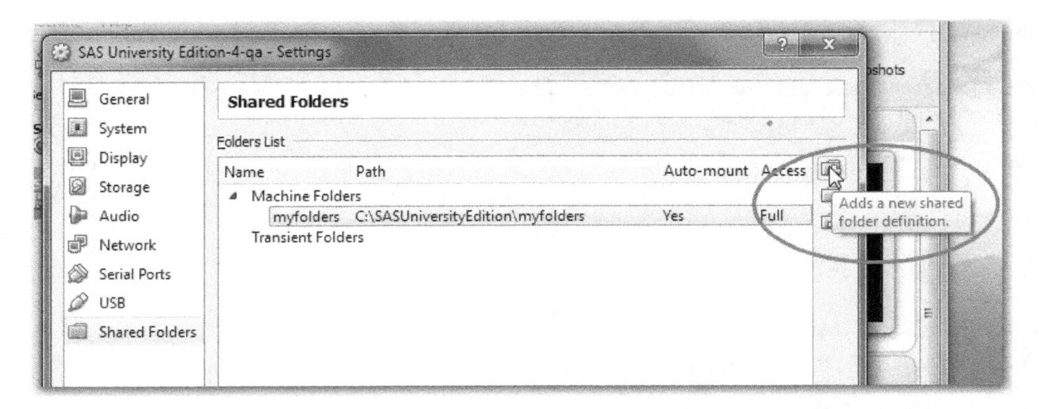

In the Folder Path pull-down list, select **Other.**

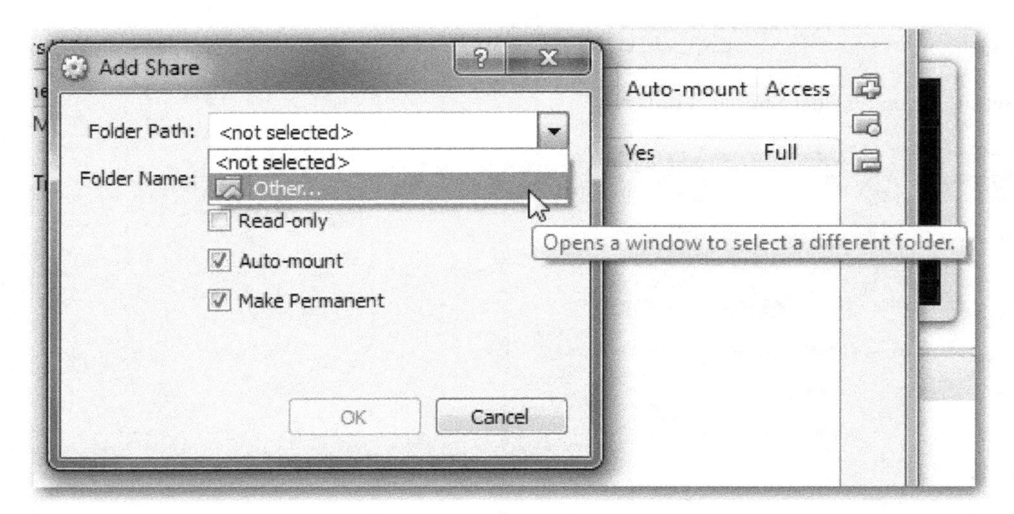

Use the **Browse** button to select the folder on your hard drive. You probably want to give this shared folder the same name as the folder name.

Note: The name of the shared folder is case sensitive and should not include any spaces or special characters.

Make sure that the two boxes labeled "Auto-mount" and "Make Permanent" are checked. Note: Depending on how you choose your folder, the selection "Make Permanent" may not be listed. This is OK.

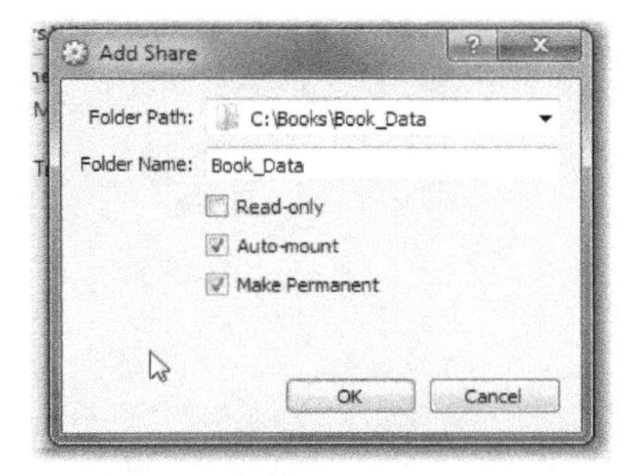

Click OK and the folder name is automatically created. Your new folder should now show up in the list of folders on your virtual machine.

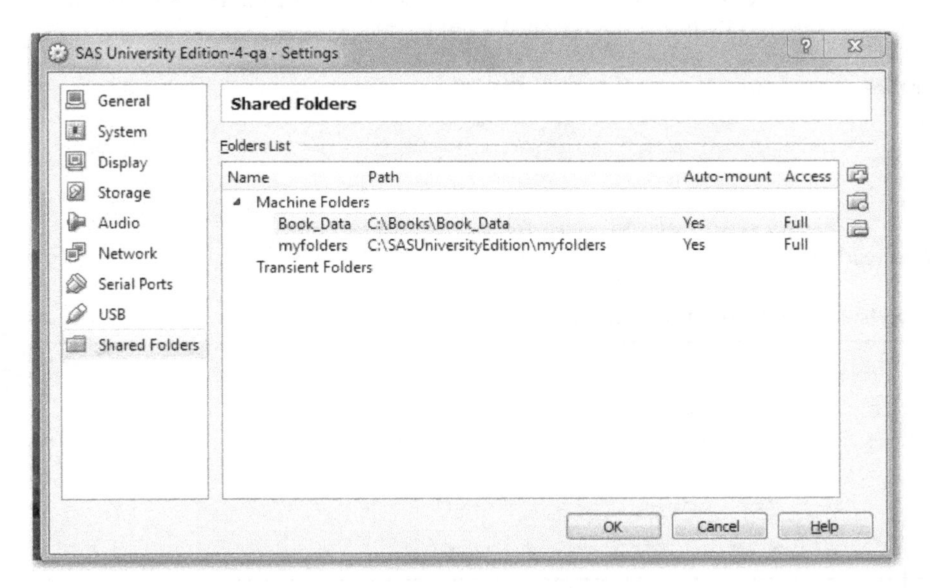

You are now finished with creating the shared folder on your virtual machine. You can add these shortcuts before you start SAS Studio or while SAS Studio is running. If you add these shortcuts while

SAS Studio is running, you must refresh the page in order for the folder shortcuts to appear in the navigation pane. The folder `sasdata` now shows up under the **Folder Shortcuts** tab as shown here:

Figure 9.3: Folder Shortcuts

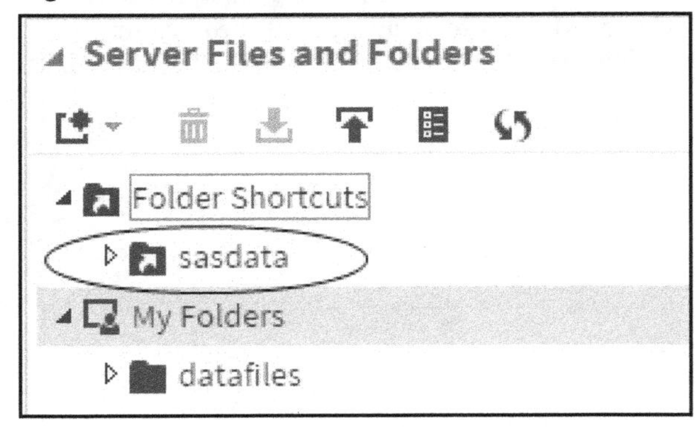

You can now access files from this folder in the folder shortcuts section in SAS Studio or you can reference this folder shortcut in a SAS program by using `/folders/myshortcuts/sasdata`.

To demonstrate how to read data from this shared folder, a file called HeightWeight.txt was created in the location `c:\sasdata`. Here is a listing of the HeightWeight.txt file:

Listing of the HeightWeight.txt file

```
59 120
65 220
62 130
45 89
```

Here is the program to read this file:

Program 9.2: Reading the HeightWeight.txt File Located in c:\sasdata

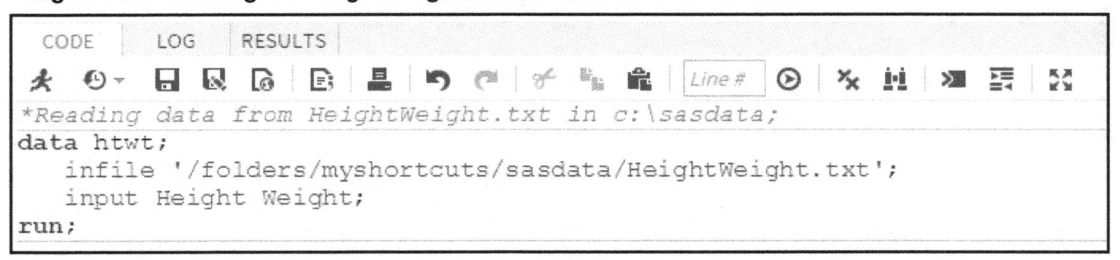

```
*Reading data from HeightWeight.txt in c:\sasdata;
data htwt;
   infile '/folders/myshortcuts/sasdata/HeightWeight.txt';
   input Height Weight;
run;
```

Notice the location on the INFILE statement. You can follow these steps to read and write data from any location on your hard drive. Luckily, you have to create the shared folder and the folder shortcut only once.

Creating a Permanent SAS Data Set

In the example that follows, you first create a permanent SAS data set (using the default `c:\SASUniversityEdition\myfolders` location). You then start a new SAS session where you read the previously created SAS data set and perform some calculations.

Here is the program to create a permanent SAS data set called Demo:

Program 9.3: Program to Create a Permanent SAS Data Set

```
*Program to create a permanent SAS data set called Demo in the
 c:\SASUniversity\myfolders folder;

libname mydata '/folders/myfolders';

data mydata.Demo;
   infile "/folders/myfolders/demographic.txt/";
   input ID $ Gender $ Age Height Weight Party $;
run;
```

The last part of the SAS log (below) shows that the data set Mydata.Demo was created.

```
NOTE: The data set MYDATA.DEMO has 5 observations and 6 variables.
NOTE: DATA statement used (Total process time):
      real time           0.03 seconds
      cpu time            0.02 seconds
```

If you look at the files under the **My Folders** tab, you see the data set listed:

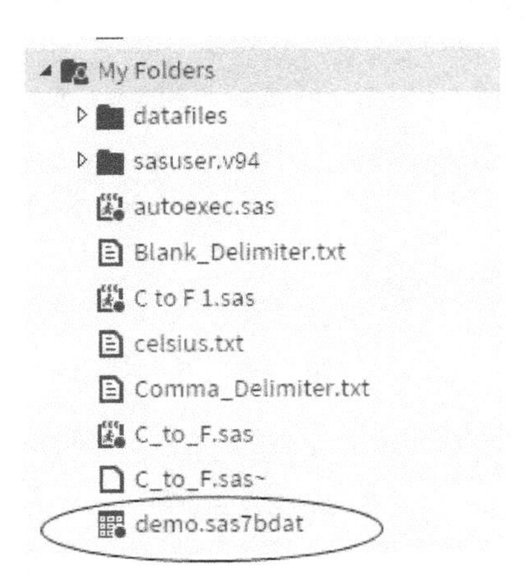

If you list the contents of `c:\SASUniversityEdition\myfolders` on your hard drive, you also see the SAS data set (below):

Name	Date modified	Type
celsius.txt	12/16/2014 5:05 PM	Text Document
Date_Data.txt	2/4/2015 10:22 AM	Text Document
demo.sas7bdat	1/12/2015 10:20 AM	SAS Data Set
demographic.txt	12/26/2014 2:19 PM	Text Document
Demographics.sas	12/27/2014 10:31 ...	SAS File
formats.sas7bcat	1/19/2015 10:29 AM	SAS Catalog
health.txt	1/22/2015 9:21 AM	Text Document

Reading from a Permanent SAS Data Set

You use an INPUT statement to read data from a raw data file—you use a SET statement to read data from a SAS data set. In the example that follows, you want to use the SAS data set Demo as input to a DATA step that creates a new variable called Wt_Kg, representing the person's weight in kilograms. In addition, let's assume that you are writing this program in a new SAS session. Here is the program:

Program 9.4: Using a SAS Data Set as Input to a Program

```
libname mydata '/folders/myfolders';

data New_Demo;
   set mydata.Demo;
   Wt_Kg = Weight / 2.2;
run;

title "Listing of Data Set New_Demo";
proc print data=New_Demo;
run;
```

In this example, you plan to create a new, temporary data set called New_Demo. Because the SAS data set Demo is stored in a folder called `myfolders`, you issue a LIBNAME statement pointing to that folder. Next, you use a SET statement to read observations from the Demo data set. You use an assignment statement to create the new variable Wt_Kg.

Here is the output:

Output from Program 9.4

Listing of Data Set New_Demo

Obs	ID	Gender	Age	Height	Weight	Party	Wt_Kg
1	012345	F	45	65	155	I	70.455
2	131313	M	28	70	220	R	100.000
3	987654	F	35	68	180	R	81.818
4	555555	M	64	72	165	D	75.000
5	172727	F	29	62	102	I	46.364

If you want the new data set New_Demo to be permanent, replace the DATA statement with:

```
data mydata.New_Demo;
```

Conclusion

It takes a while to get comfortable reading and writing permanent SAS data sets using the SAS University Edition. If you use the folder location suggested in setting up SAS University Edition for the first time (`\SASUniversityEdition\myfolders`), SAS Studio can read and write to this location without having to create additional shared folders on your virtual machine. As you are learning how to write SAS programs, it is advisable to use this default folder.

Problems

1. Use PROC CONTENTS to display the data descriptor for the data set Heart in the Sashelp library. Run it again with the VARNUM procedure option (remember, you place procedure options between the procedure name and the semicolon).
2. Create a new, temporary SAS data set called Heart_Vars from the data set Heart in the Sashelp library. Include the variables BP_Status, Chol_Status, Systolic, Diastolic, and Status. Use a KEEP= data set option on the Heart data set to do this. Hint: First, click on the **My Libraries** tab and then expand the list of the data sets in Sashelp. Next, hold the Ctrl key down and select the requested variables in the order they are listed in this problem. Finally, right-click on any of the marked files and then drag the list to your code.
3. Repeat Problem 2 except make the data set Heart_Vars a permanent data set in your `myfolders` folder.
4. Create a temporary SAS data set called Alive from the data set Heart in the Sashelp library. This data set should contain the variables BP_Status, Chol_Status, Systolic, and Diastolic. Use a WHERE= data set option to select only those observations where Status is equal to 'Alive'. Use the data set option (OBS=10) with PROC PRINT to list the Alive data set, like this:

   ```
   proc print data=Alive(obs=10);
   ```

   Hint: You will need to include the variable Status in the KEEP= data set option because you need this variable to use in your WHERE= data set option. Use a DROP statement to drop Status so that it is not included in the Heart_Vars data set.
5. (Advanced) Create a folder on your hard drive. Open your virtual machine and create a shared folder called `Sasdata` associated with this location. Open SAS Studio and write the statements necessary to create a permanent SAS data set called Young_Males in the Sasdata library. Use as input the Sashelp data set Class and select only those observations where Gender is equal to 'M' and Age is 11 or 12.
6. What's wrong with this program?

   ```
   1. data New;
   2.    set SASHELP.Fish(keep Species Weight);
   3.    Wt_Kg = Weight/2.2;
   4.    *Note: 1 Kg = 2.2 Lbs *;
   5. run;
   ```

Chapter 10: Creating Formats and Labels

What Is a SAS Format and Why Is It Useful?

It is a common practice to store information in a database using codes rather than actual values. For example, you might have a questionnaire where the responses are strongly disagree, disagree, neutral, agree, and strongly agree. It would be unusual to store the actual values in your database—rather, you would use coded values such as 1=strongly disagree, 2=disagree, and so on.

Even though you are storing codes in your database, you would like to see the actual labels printed in your output. SAS formats are the tool that allows this to happen.

To demonstrate how to create your own SAS formats, let's start with a SAS data set called Taxes, described in the table below:

Variable	Description	Codes Used
SSN	Social Security Number	
Gender	Gender	M=Male F=Female

Variable	Description	Codes Used
Question_1	Do you pay taxes?	1=Yes, 0=No
Question_2	Are you satisfied with this service?	1=strongly disagree, 2=disagree, 3=neutral, 4=agree, 5=strongly agree
Question_3	How many phone calls did it take to resolve your problem?	A=0, B=1 or 2, C=3 to 5, D=More than 5
Question_4	Was the person answering your call friendly?	Same as Question_2
Question_5	How much did you pay?	The actual dollar amount

The program to create the Taxes data set is shown next:

Program 10.1: Creating the Taxes Data Set (and Demonstrating a DATALINES Statement)

```
data Taxes;
   informat SSN $11.
            Gender $1.
            Question_1 - Question_4 $1.;

    input SSN Gender Question_1 - Question_5;
datalines;
101-23-1928 M 1 3 C 4 23000
919-67-7800 F 9 2 D 2 17000
202-22-3848 M 0 5 A 5 57000
344-87-8737 M 1 1 B 2 34123
444-38-2837 F . 4 A 1 17233
763-01-0123 F 0 4 A 4 .
;

title "Listing of Data Set TAXES";
proc print data=Taxes;
   id SSN;
run;
```

This program uses a feature you have not seen before: a DATALINES statement. When you want to write a short SAS program to test your logic or syntax, you can save the trouble of writing a text file (perhaps using Notepad or some similar editor) and then writing an INFILE statement telling the program where to find the data. Instead, you can actually include the lines of data in the program itself. You do this by writing a DATALINES statement and following this statement with your lines of data. You end your data with a single semicolon or a RUN statement. SAS will read these lines of data as if they were in an external file. There is one additional new feature in this program—that is the ID statement following PROC PRINT. You use an ID statement to do two things: First, you specify a variable that you want to display in the first column of your listing. If your data set has more variables that can fit across one page of output, the ID variable is repeated on each new page. Second, when you specify an ID variable, the default Obs column is no longer printed (this is a good thing in this author's opinion).

Here is a listing of the Taxes data set:

Listing of Data Set TAXES

SSN	Gender	Question_1	Question_2	Question_3	Question_4	Question_5
101-23-1928	M	1	3	C	4	23000
919-67-7800	F	9	2	D	2	17000
202-22-3848	M	0	5	A	5	57000
344-87-8737	M	1	1	B	2	34123
444-38-2837	F		4	A	1	17233
763-01-0123	F	0	4	A	4	.

You would prefer to have values for Gender be listed as 'Male' and 'Female', values for Question_1 to be listed as 'Yes' and 'No', values for Question_2 and Question_4 to display the agreement scale (called a *Likert scale* by psychometricians), and values for Question_3 to show the number of calls. Finally, you would like to place the dollar amounts into four categories: 0 - $10,000 = Low, $10,001 - $20,000 = Medium, $20,001 - $50,000 = High, and $50,000+ = Very High.

The process of substituting these labels for the coded values is called *formatting* in SAS terminology. The first step is to create the formats. Once that is accomplished, you can associate one or more variables with these formats. You use PROC FORMAT to create your SAS formats, as shown in the following program:

Program 10.2: Creating Your Own Formats

```
proc format;
   value $Gender 'M'='Male'
                 'F'='Female';
   value $Yesno '0'='No'
                '1'='Yes'
           other='Did not answer';
   value $Likert 1='Strongly Disagree'
                 2='Disagree'
                 3='No Opinion'
                 4='Agree'
                 5='Strongly Agree';
   value $Calls 'A'='None'
                'B'='1 or 2'
```

```
                   'C'='3 - 5'
                   'D'='More than 5';
   value Pay_group  low-10000 = 'Low'
                    10001-20000 = 'Medium'
                    20001-50000 = 'High'
                    50001-high = 'Very High';
run;
```

You write VALUE statements to define your formats and follow the keyword VALUE with the name of the format you wish to create. Format names can be up to 32 characters long, and they follow the same naming conventions as other SAS names, except that they cannot end in a digit. Formats that you plan to use with character formats must begin with a dollar sign ($) (leaving 31 characters for you to use). In this example, the first four formats will be associated with one or more character variables. The last format will be associated with the numeric variable Question_5. Because the first four formats will be associated with character variables, they all begin with a dollar sign. Following the format name, you list either a single value (such as 'M') or a range of values (such as low–10000). Notice that you place the character values in single or double quotes. There are several keywords that you can use in defining a value. As you can see in the $Yesno format, the keyword OTHER will supply the label "Did not answer" for all values other than '0' or '1'. You can also use the keywords LOW and HIGH to refer to the lowest and highest values, respectively. Note that for the Pay_group format, you cannot include commas in the numerical ranges.

To demonstrate how formats work, let's first run PROC FREQ to compute frequencies for each variable in the Taxes data set, without formatting any of the variables. The PROC FREQ statements look like this:

Program 10.3: Computing Frequencies on the Taxes Data Set (without Formats)

```
title 'Frequencies for the Taxes Data Set';
proc freq data=Taxes;
   tables Gender Question_1 - Question_5 / nocum;
run;
```

The TABLES option NOCUM is an instruction to omit cumulative frequencies from the results as shown below:

Output from Program 10.3

The FREQ Procedure

Gender	Frequency	Percent
F	3	50.00
M	3	50.00

Question_1	Frequency	Percent
0	2	40.00
1	2	40.00
9	1	20.00
Frequency Missing = 1		

Question_2	Frequency	Percent
1	1	16.67
2	1	16.67
3	1	16.67
4	2	33.33
5	1	16.67

Question_3	Frequency	Percent
A	3	50.00
B	1	16.67
C	1	16.67
D	1	16.67

Question_4	Frequency	Percent
1	1	16.67
2	2	33.33
4	2	33.33
5	1	16.67

Question_5	Frequency	Percent
17000	1	20.00
17233	1	20.00
23000	1	20.00
34123	1	20.00
57000	1	20.00
Frequency Missing = 1		

As you can see, PROC FREQ listed frequencies for each unique value of the variables. For numeric variables such as Question_5 (How much did you pay, in dollars?), you see frequencies for each unique value—not something that is very useful. There are other SAS procedures (such as PROC SGPLOT) that will automatically place numerical values into groups and produce histograms, etc.

It's time to see what the output from PROC FREQ looks like when you add a FORMAT statement, associating each of the variables in the Taxes data set with a format. Here is the code:

Program 10.4: Adding a Format Statement to PROC FREQ

```
title 'Frequencies for the Taxes Data Set';
proc freq data=Taxes;
   format Gender $Gender.
          Question_1 $Yesno.
          Question_2 Question_4 $Likert.
          Question_3 $Calls.
          Question_5 Pay_group.;
   tables Gender Question_1 - Question_5 / nocum;
run;
```

You use a FORMAT statement to associate your variables with the appropriate format. SAS programs recognize the difference between variables and formats because you end each format name with a period. The two variables Question_2 and Question_4 share the same format, so you list these two variables together and follow them with the format you want to use ($Likert). Output from this program is shown below:

Output from Program 10.4

Frequencies for the Taxes Data Set

The FREQ Procedure

Gender	Frequency	Percent
Female	3	50.00
Male	3	50.00

Question_1	Frequency	Percent
No	2	50.00
Yes	2	50.00
Frequency Missing = 2		

Question_2	Frequency	Percent
Strongly Disagree	1	16.67
Disagree	1	16.67
No Opinion	1	16.67
Agree	2	33.33
Strongly Agree	1	16.67

Question_3	Frequency	Percent
None	3	50.00
1 or 2	1	16.67
3 - 5	1	16.67
More than 5	1	16.67

Question_4	Frequency	Percent
Strongly Disagree	1	16.67
Disagree	2	33.33
Agree	2	33.33
Strongly Agree	1	16.67

Question_5	Frequency	Percent
Medium	2	40.00
High	2	40.00
Very High	1	20.00
Frequency Missing = 1		

You now see the formatted values in the tables. You may wonder about the order of the values in the tables. Notice that Female comes before Male and No comes before Yes. Did you guess that PROC FREQ orders these values alphabetically (by their original internal values)? If so, you are correct. Later on, you will see how to change the order in the frequency tables.

If you are very observant, you will have noticed that the table for Question_1 lists two missing values. However, when you look at the unformatted data, you see only one missing value and one value of 9. It turns out that the OTHER category was combined with the missing values (because it was not a valid value for this variable). If you want to separate the missing value from the OTHER category, you can add a format for a missing value like this:

```
value $Yesno '0'='No'
             '1'='Yes'
             ' '='Did not answer'
             other='Invalid value';
```

Because $Yesno is a character format, you specify a missing value with a single space between the quotes. To specify a missing value for a numeric format, use a single period instead. When you use this format with PROC FREQ, the output frequencies for Question_1 look like this:

Using the Modified $Yesno Format

Frequencies for Question 1

The FREQ Procedure

Question_1	Frequency	Percent
No	2	40.00
Yes	2	40.00
Invalid value	1	20.00
Frequency Missing = 1		

By using this modified format, you now see that there was one invalid value and one missing value.

Using SAS Built-in Formats

SAS provides you with a large number of built-in formats that you can use, along with ones you create yourself. One example of a numeric format is *w.d*, where *w* (stands for width) specifies the total number of spaces to use when writing a number and *d* specifies how many places to include after the decimal place. Here are some examples:

For each of these examples, X=1234.567

Format	Display	Explanation
8.3	1234.567	The 8 is the total column width, including the decimal point.
10.4	1234.5670	There is a leading space and a 0 is added to give four places to the right of the decimal point.
8.2	1234.57	There is one leading blank and the decimal value is rounded.
4.	1235	When the width is shorter than the value, it gets truncated (and rounded).
10.1	1234.6	There are four leading blanks and the decimal is rounded.

Some other useful SAS formats are dollar*w.d* and comma*w.d.* The dollar format adds dollar signs, commas (if needed), and cents (if *d* is equal to 2). The value of *w* is the field width. This includes the dollar sign, any commas, a decimal point, and the digits to the right of the decimal point. If you leave out a value of *d* (the number after the period), the dollar value will be rounded. The comma format is similar to the dollar format, except that it does not include a dollar sign and the value of *d* can show as many decimal places as needed. Formats for SAS dates are particularly useful and will be discussed in the chapter on SAS dates.

More Examples to Demonstrate How to Write Formats

There is great flexibility in defining values or ranges when you create a SAS format. The following examples help illustrate this flexibility.

In Program 10.1, one of the formats placed numerical values into four categories as shown here:

```
value Pay_group  low-10000   = 'Low'
                 10001-20000 = 'Medium'
                 20001-50000 = 'High'
                 50001-high  = 'Very High';
```

This works fine as long as the values to be formatted are integers. However, suppose you tried to format a value of 10,000.50. This value falls between the Low and Medium ranges (and would not be formatted—it would print as 10000.50 in a listing). You might specify ranges like this:

```
value Pay_group  low-10000    = 'Low'
                 10000-20000 = 'Medium'
                 20000-50000 = 'High'
                 50001-high  = 'Very High';
```

Although this code works, it is confusing. What is the formatted value of 10,000? It turns out it would be formatted as 'Low'. It is better to allow a value to only match a single range. You can exclude a value from the beginning or the end of a range by adding a less than (<) sign before or after the dash that specifies ranges. As an example, to exclude 10,000 from the low range, you would use:

```
value Pay_group  low-<10000 = 'Low'
```

If you wanted to exclude 10,000 from the medium range, you would use:

```
value Pay_group  10000<-20000 = 'Medium'
```

You can also use dashes and commas to specify character ranges. Here are some examples:

```
value $Grades 'A' ,'B'  = 'Good'
              'C' - 'E' = 'Passing'
              'F'       = 'Fail'
              other     = 'Error';
```

Values of A or B are formatted as 'Good'; C, D, and E are formatted as 'Passing'; F is formatted as 'Fail'; and any characters not equal to any of these values is formatted as 'Error'.

Describing the Difference between a FORMAT Statement in a Procedure and a FORMAT Statement in a DATA Step

In the programs you have seen thus far, the FORMAT statements have been placed inside a SAS procedure. This creates an association between the variables and formats only for that procedure. If you place a FORMAT statement in a DATA step, the association between the variables and formats remains for the entire program. If you write a PROC PRINT or PROC FREQ step, you will see formatted values for all the variables you listed in your FORMAT statement.

You will usually find it more convenient to write your FORMAT statement in a DATA step, saving the trouble of having to rewrite (or copy) it for each procedure you run. Keep in mind that even though you have assigned a format to a variable, many procedures such as PROC MEANS or other statistical procedures will still use the internal values of the variables when doing their calculations.

Making Your Formats Permanent

All of the formats previously described are temporary formats—that is, they only exist for the duration of your SAS session. If you have formats that you plan to use frequently, you can make them permanent. That way, every time you open up a SAS session, you can use any of your permanent formats without having to rerun PROC FORMAT. There are a few steps you need to follow to make this happen.

First, you need to decide where you plan to store your formats. To keep this example simple, let's store your permanent formats in `myfolders`. You run PROC FORMAT just as you did previously, except you add the procedure option LIBRARY=*libref* (where *libref* is the library reference you create using a LIBNAME statement). Here is an example:

Program 10.5: Making a Permanent Format

```
libname myfmts '/folders/myfolders';

proc format library=myfmts;
   value $Gender 1='Male'
                 2='Female';
   value $Yesno  0='No'
                 1='Yes';
run;
```

Here is a copy of the SAS log after you run this program:

```
39         libname myfmts '/folders/myfolders';
NOTE: Libref MYFMTS was successfully assigned as follows:
      Engine:        V9
      Physical Name: /folders/myfolders
40         proc format library=myfmts;
41            value $Gender 1='Male'
42                          2='Female';
NOTE: Format $GENDER has been written to MYFMTS.FORMATS.
43            value $Yesno  0='No'
44                          1='Yes';
NOTE: Format $YESNO has been written to MYFMTS.FORMATS.
45         run;
```

The two formats, $Gender and $Yesno, are now permanent formats and are stored in the Myfmts library. If you use your operating system to list the files in the folder called `SASUniversityEdition/myfolders`, you will see a file called formats.sas7bcat (the file extension shows that the formats are compatible with SAS 7 and they are stored in a binary catalog).

If you have associated formats with a permanent SAS data set, keep in mind that the formats are now a permanent property of those variables, and the format definitions (catalog) must be available in order to open and read the data set. That is, if you give someone a copy of one of your permanent SAS data sets, be sure to give them the format catalog (it will have the name formats.sas7bcat).

A good way to keep track of your permanent formats is to include another PROC FORMAT option called FMTLIB. This option produces a listing of all the formats in the specified library, along with all of the defined values for these formats. To illustrate this, let's run Program 10.5 again, with the FMTLIB option added.

Program 10.6: Adding the FMTLIB Option

```
libname myfmts '/folders/myfolders';

proc format library=myfmts fmtlib;
   value $Gender 1='Male'
                 2='Female';
   value $Yesno  0='No'
                 1='Yes';
run;
```

Here is the output:

Output from Program 10.6

```
----------------------------------------------------------------------------
|       FORMAT NAME: $GENDER  LENGTH:    6   NUMBER OF VALUES:    2        |
|   MIN LENGTH:   1  MAX LENGTH:  40  DEFAULT LENGTH:   6  FUZZ:        0  |
|--------------------------------------------------------------------------|
|START           |END             |LABEL  (VER. V7|V8   19JAN2015:11:29:26)|
|----------------+----------------+----------------------------------------|
|1               |1               |Male                                    |
|2               |2               |Female                                  |
----------------------------------------------------------------------------


----------------------------------------------------------------------------
|       FORMAT NAME: $YESNO   LENGTH:    3   NUMBER OF VALUES:    2        |
|   MIN LENGTH:   1  MAX LENGTH:  40  DEFAULT LENGTH:   3  FUZZ:        0  |
|--------------------------------------------------------------------------|
|START           |END             |LABEL  (VER. V7|V8   19JAN2015:11:29:26)|
|----------------+----------------+----------------------------------------|
|0               |0               |No                                      |
|1               |1               |Yes                                     |
----------------------------------------------------------------------------
```

This output is a very useful document for you to use or to share with others who wish to use your formats themselves.

Before you use these permanent formats in a new SAS session, you need to include two statements: One is the OPTIONS FMTSEARCH=*libref* statement. This option tells SAS to look in the location specified by your *libref* to find your permanent formats. The other statement redefines your *libref*. So, beginning every SAS session, you would add these two statements:

```
libname myfmts '/folders/myfolders';
options fmtsearch=(myfmts);
```

This is a good time to tell you about a special file called Autoexec.sas. This is a file where you can place SAS statements and options. All the statements and options placed in the Autoexec.sas file execute every time you open a SAS session. To create your Autoexec.sas file, go to the SAS Studio menu, and select **Edit Autoexec File**, as shown in the figure below:

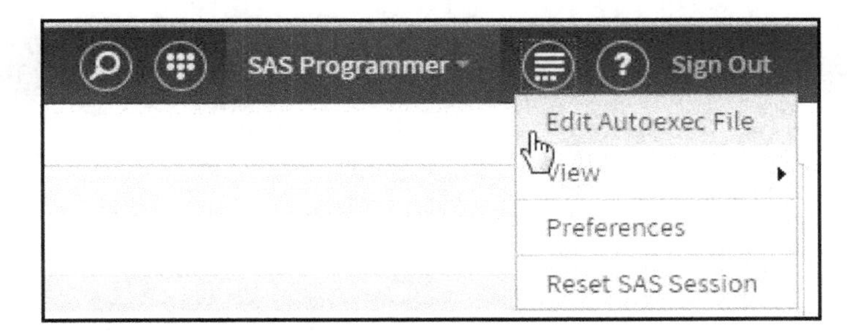

Now you can type in your favorite options and other statements that you want submitted at the beginning of each SAS session. Here is an example:

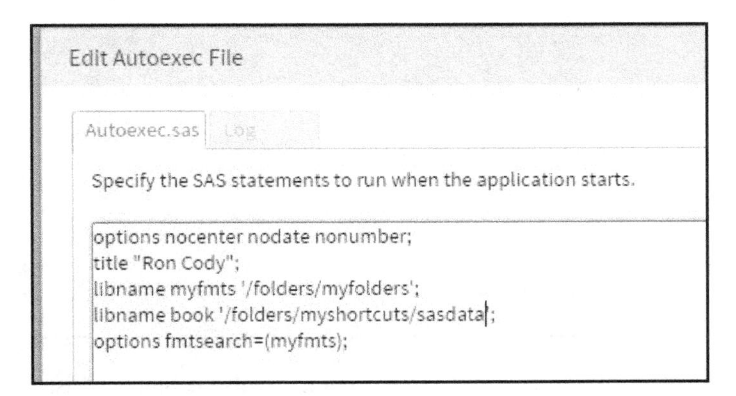

In this Autoexec.sas file, I have set some options, supplied a default title, created two library references (to the same folder), and added the FMTSEARCH= system option. No doubt, you will want to use a different TITLE statement.

Creating Variable Labels

You can associate labels with your SAS variables. If you have variables such as Gender and Race, a label may not be necessary. However, for variable names such as Question_1, Question_2, etc., you might want to provide labels. You use a LABEL statement to associate labels with your variables. As an example, the following program adds labels to the variables in the Taxes data set (described in the beginning of this chapter):

Program 10.7: Creating Variable Labels

```
data Taxes;
   informat SSN $11.
            Gender $1.
            Question_1 - Question_4 $1.;

    input SSN Gender Question_1 - Question_5;
    label Question_1 = 'Do you pay taxes?'
          Question_2 = 'Are you satisfied with the service?'
          Question_3 - 'How many phone calls?'
          Question_4 = 'Was the person friendly?'
          Question_5 = 'How much did you pay?';

datalines;
101-23-1928 M 1 3 C 4 23000
919-67-7800 F 9 2 D 2 17000
202-22-3848 M 0 5 A 5 57000
344-87-8737 M 1 1 B 2 34123
444-38-2837 F . 4 A 1 17233
763-01-0123 F 0 4 A 4 .
;

title 'Frequencies for the Taxes Data Set';
proc freq data=Taxes;
   format Gender $Gender.
          Question_1 $Yesno.
          Question_2 Question_4 $Likert.
          Question_3 $Calls.
          Question_5 Pay_group.;
   tables Gender Question_1 - Question_5 / nocum;
run;
```

Now that you have added variable labels to your program, let's see how it affects the output. Listed below is a partial output from this program:

Output from Program 10.7

Do you pay taxes?		
Question_1	Frequency	Percent
0	2	40.00
1	2	40.00
9	1	20.00
Frequency Missing = 1		

Are you satisfied with the service?		
Question_2	Frequency	Percent
1	1	16.67
2	1	16.67
3	1	16.67
4	2	33.33
5	1	16.67

You now see the variable label listed at the top of each frequency table.

Conclusion

Adding formats and labels to a SAS program will make the listings and tables much more readable. If you have formats that you use frequently, be sure to create permanent formats so that you don't have to run PROC FORMAT every time you start a SAS session. If you have associated formats with variables in the DATA step, and plan to share your data set with others, be sure to include the format catalog (formats.sas7bcat) along with the SAS data set.

Problems

1. Modify the following program to supply formats to the variables listed in the table below:

Variable	Type	Formatted Values
Gender	Num	1=Male, 2=Female
Q1-Q4	Char	1='Strongly Disagree', 2='Disagree', 3='No Opinion', 4='Agree', 5='Strongly Agree'

Variable	Type	Formatted Values
Visit	Num	Month/Day/Year
Age	Num	0-20='Young', 21=40='Still young', 40-60='Middle', 61+='Older'

Program for Problem Sets 1

```
data Questionnaire;
   informat Gender 1. Q1-Q4 $1. Visit date9.;
   input Gender Q1-Q4 Visit Age;
   format Visit date9.;
datalines;
1 3 4 1 2 29May2015 16
1 5 5 4 3 01Sep2015 25
2 2 2 1 3 04Jul2014 45
2 3 3 3 4 07Feb2015 65
;
title "Listing of Data Set QUESTIONNAIRE";
proc print data=Questionnaire noobs;
run;
```

2. Using the program in Problem 1, create a format that places ages into the following categories:

   0-20='Group 1', 21-40='Group 2', 41-60='Group 3', 61-80='Group 5', 81+='Group 5'

   Use this format for the variable Age and the other formats described in Problem 1. Produce a listing showing the formatted values for all the variables.

3. You have a character variable called Grades. Values of Grades are 'A', 'B', 'C', 'D', 'F', 'I', and missing. Write a format (call it $Grades) that formats 'A' and 'B' as 'Good', 'C' as 'Average', 'D' as 'Poor', 'F' as 'Fail', 'I' as 'Incomplete', and missing values as 'Missing'. Also, include a format for any nonmissing value that is not one of the valid values (call them Invalid).

Chapter 11: Performing Conditional Processing

Introduction

All programming languages allow you to perform *conditional processing*—making decisions based on data values or other conditions. As an example, you might want to create a new variable (Age_Group) based on the values of age. Another common use of conditional logic is to check if data values are within a prescribed range. Performing operations of this type requires conditional processing.

Grouping Age Using Conditional Processing

For the first example, you have data on gender, age, height, and weight. You want to create a new variable (Age_Group) based on the variable Age. Here is a first attempt that runs but has a logical flaw in regard to SAS missing values:

Program 11.1: First Attempt at Creating an Age Group Variable (Incorrect Program)

```
data People;
   input @1  ID      $3.
         @4  Gender $1.
         @5   Age     3.
         @8  Height  2.
         @10 Weight  3.;
```

```
   if Age le 20 then Age_Group = 1;
   else if Age le 40 then Age_Group = 2;
   else if Age le 60 then Age_Group = 3;
   else if Age le 80 then Age_Group = 4;
   else if Age ge 80 then Age_Group = 5;

datalines;
001M 5465220
002F10161 98
003M 1770201
004M 2569166
005F   64187
006F 3567135
;

title "Listing of Data Set PEOPLE";
proc print data=People;
   id ID;
run;
```

To indicate conditions such as less than, etc., you have a choice of two-letter abbreviations or symbols. The table below shows all the possible logical comparisons:

Logical Comparison Operators

Logical Comparison	Mnemonic	Symbol
Equal to	EQ	=
Not equal to	NE	^= or ~= or ¬=
Less than	LT	<
Less than or equal to	LE	<=
Greater than	GT	>
Greater than or equal to	GE	>=
Equal to any value in a list	IN	

The new statements in this program are the IF and ELSE IF statements. They work like this: Following the IF or ELSE IF statement is a logical statement that is either true or false. If the statement is true, the following expression executes; if it is false, the following expression does not execute. Also, if the logical statement on an IF or ELSE IF statement is true, all the subsequent ELSE IF statements are skipped. For example, in data set People, the first subject is 54 years old. The first ELSE IF statement that is true is

```
else if age le 60 then Age_Group = 3;
```

Because this statement is true, all the remaining ELSE IF statements are skipped. This logic has the advantage of being more efficient than a series of IF statements—the program does not have to evaluate more IF statements than necessary.

Let's run the program and examine the output:

Output from Program 11.1

Listing of Data Set PEOPLE

ID	Gender	Age	Height	Weight	Age_Group
001	M	54	65	220	3
002	F	101	61	98	5
003	M	17	70	201	1
004	M	25	69	166	2
005	F	.	64	187	1
006	F	35	67	135	2

Most of the Age_Group values are correct. However, there is a problem for ID 005. This person had a missing value for age but was placed in age group 1. Why? In SAS, a numeric missing value is treated logically as the most negative number possible. Thus, a missing value is less than any real—positive or negative—number. The first IF statement asks if Age is less than or equal to 20. Person 005 has a missing value for Age and a missing value is less than 20 so this person is placed in age group 1. Here is one way to fix Program 11.1:

Program 11.2: Corrected Version of Program 11.1

```
data People;
   input @1   ID      $3.
         @4   Gender $1.
         @5    Age     3.
         @8   Height   2.
         @10 Weight   3.;
   if missing(Age) then Age_Group = .;
   else if Age le 20 then Age_Group = 1;
   else if Age le 40 then Age_Group = 2;
   else if Age le 60 then Age_Group = 3;
   else if Age le 80 then Age_Group = 4;
   else if Age ge 80 then Age_Group = 5;
```

```
datalines;
001M 5465220
002F10161 98
003M 1770201
004M 2569166
005F   64187
006F 3567135
;

title "Listing of Data Set PEOPLE";
proc print data=People;
   id ID;
run;
```

The first IF statement tests if Age is a missing value. This is accomplished using the MISSING function. All SAS functions contain a set of parentheses. The values placed in the parentheses are called *arguments* to the function. The MISSING function returns a value of true if the argument is a missing value and false otherwise (this works for both character and numeric arguments). When the program processes ID 005, the missing function returns a true value and the variable Age_Group is set to a missing value (designated by a period). An alternative to testing for a missing value is the following line of code:

```
if Age = . then Age_Group = . ;
```

This author strongly recommends that you use the MISSING function to test for missing values.

> Most SAS programmers would agree that failing to account for missing values in a DATA step is the most common logical error in SAS programming. Always be sure that you consider the consequences of a missing value meeting your program logic.

Output from Program 11.2 results in a missing value for ID 005.

Using Conditional Logic to Check for Data Errors

You can use IF-THEN-ELSE logic to test if values of certain variables are outside a predetermined range. For example, you might want to check if anyone in the People data set was heavier than 200 pounds or lighter than 100 pounds. The following program does just that:

Program 11.3: Using Conditional Logic to Test for Out-of-Range Data Values

```
data _null_;
   set People;
   if Weight lt 100 and not missing(Weight) or Weight gt 200 then
   put "Weight for ID " ID "is " Weight;
run;
```

There are several new features in this program. First is the special data set name _NULL_. This is a reserved data set name that allows you to run a DATA step without actually creating a data set. The reason it is used in this program is that you are checking for out-of-range values, and you do not need a data set when you are finished checking—thus the use of DATA _NULL_. Because you are not creating a data set, the program is more efficient than one that does create a data set.

The SET statement brings in observations from the People data set. Notice that the condition on the IF statement checks for two things: First, is the value less than 100 and not missing (remember that a missing value is less than 100)? Second, is the Weight greater than 200? If either condition is true, the PUT statement executes. It is an instruction to write out the text "Weight for ID" followed by the value of ID (note ID is not in parentheses so it represents a variable name) followed by the word "is" followed by the value of Weight. By default, the PUT statement writes its output to the SAS log. This is fine for programmers but not so fine for nonprogrammers. To tell SAS to write out the Weight values to the RESULTS window, add the line `file print;` before the PUT statement. Here is the output from Program 11.3:

Output from Program 11.3

```
40         data _null_;
41            set People;
42            if Weight lt 100 and not missing(Weight) or Weight gt 200 then
43            put "Weight for ID " ID "is " Weight;
44         run;
Weight for ID 001 is 220
Weight for ID 002 is 98
Weight for ID 003 is 201
```

Three people had weights out-of-range.

Describing the IN Operator

If you want to check if a value is any one of several values, you can use multiple OR operators or the IN operator. Suppose you want to check if values for Race are 'W', 'B', 'H', or 'O' (white, black, Hispanic, or other). Using the OR operator, you could write:

```
length Race_Value $ 7;
if Race = 'W' or Race = 'B' or Race = 'H' or Race = 'O' then
   Race_Value = 'Valid';
else Race_Value = 'Invalid';
```

Using the IN operator, this line of code is much simpler:

```
length Race_Value $ 7;
if Race in ('W','B','H','O') then Race_Value = 'Valid';
else Race_Value = 'Invalid';
```

You place the character values in quotes (single or double) and separate each value by a comma or space. The statement evaluates as true if Race matches any one of the listed values. It is useful to know that once a match is made, the IN operator stops looking for matches. When extreme efficiency is needed, programmers will place values most likely to be present in the data at the beginning of the list, saving the extra CPU time in searching the whole list.

You can use the IN operator with numeric data as well. For example, if you want to list all subjects in Age_Group 3, 4, or 5, you could use the following statement:

```
if Age_Group in (3,4,5) then put "Heavier Folks";
```

You may wonder about the LENGTH statement in these code segments. Why is it needed? Remember that the length of character variables is set at compile time (before any data values are read or any conditional logic is performed). Without the LENGTH statement, the first time the variable Race_Value appears is where it is set equal to 'Valid'. Because 'Valid' is 5 characters long, SAS would set the storage length for Race_Value to 5. The effect would be to truncate the other possible value for Race_Group, 'Invalid', to 5 characters. By using a LENGTH statement before the assignment statement for Race_Value, SAS assigns a storage length of 7 for this variable. A popular trick to avoid typing the LENGTH statement is to pad the first value of Race_Value, 'Valid', with two extra blanks (for example, 'Valid '), so that SAS will assign the variable a length of 7 instead of 5.

Using Boolean Logic (AND, OR, and NOT Operators)

You can combine the three Boolean operators AND, OR, and NOT in logical expressions. Here is an example:

```
*Note: there are no missing values of LDL and HDL in the data;
if (LDL gt 100 or HDL lt 50) and Gender eq 'F' then
   Risk = 'High';
   else Risk = 'Low';
if (LDL gt 100 or HDL lt 40) and Gender = 'M' then
   Risk = 'High';
   else Risk = 'Low';
```

High values of LDL (low-density lipids) or low values of HDL (high-density lipids) are considered a risk for coronary artery disease. In addition, the Mayo Clinic uses different values for the HDL cutoff for men and women.

The order of precedence of the Boolean operators in decreasing order is:

1. NOT
2. AND
3. OR

You can use parentheses to force a different order of operation. For example:

```
if x and y or z;
```

is equivalent to

```
if (x and y) or z;
```

If you want to perform the OR operation before the AND operation, write the expression like this:

```
if x and (y or z);
```

Because the NOT operator has the highest precedence, the expression:

```
if x and not y or z;
```

is equivalent to

```
if x and (not y) or z;
```

> Even though the Boolean operators have a built-in ordering, feel free to add parentheses in your logical expressions—it makes the logic easier to understand.

A Special Caution When Using Multiple OR Operators

It is very easy to make a serious error when using multiple OR operators. Take a look at the following program:

Program 11.4: A Common Error Using Multiple OR Operators

```
data Mystery;
   input x;
   if x = 3 or 4 then Match = 'Yes';
   else Match = 'No';
datalines;
3
4
9
.
-5
;
title "Listing of Data Set MYDERY";

proc print data=Mystery noobs;
run;
```

Before you look at the output, notice that the PROC PRINT option NOOBS was added. This option removes the Obs column from the output of PROC PRINT. If you use an ID statement with PROC PRINT, the NOOBS option is not necessary because the ID variable(s) replace the Obs column.

Here is the output:

Output from Program 11.4

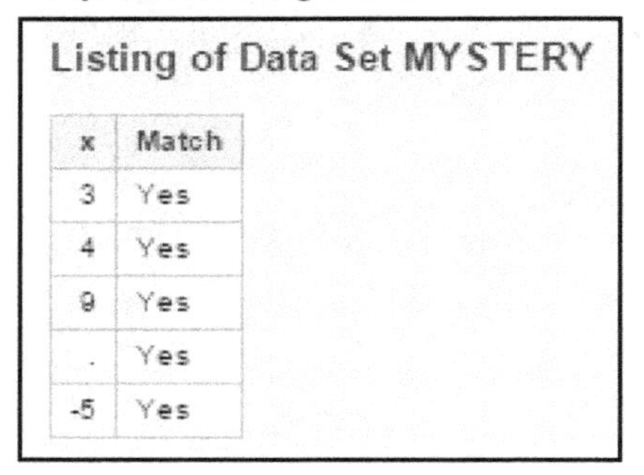

Listing of Data Set MYSTERY

x	Match
3	Yes
4	Yes
0	Yes
.	Yes
-5	Yes

That is not what you expected, is it? First of all, you probably expected a syntax error because the statement:

```
if x = 3 or 4 then Match = "yes';
```

should have been written as:

```
if x = 3 or x = 4 then Match = 'Yes';
```

However, if you expressed the logic in this program verbally, you might very well say if *x* equals 3 or 4, then Match equals 'Yes'. Let's first look at the SAS log.

SAS Log from Program 11.4

```
39          data Mystery;
40             input x;
41             if x = 3 or 4 then Match = 'Yes';
42             else Match = 'No';
43          datalines;

NOTE: The data set WORK.MYSTERY has 5 observations and 2 variables.
NOTE: DATA statement used (Total process time):
      real time           0.00 seconds
      cpu time            0.01 seconds
```

There are no syntax errors, yet the output is obviously wrong. What is going on?

> In SAS, any numerical value that is not equal to 0 or a missing value is considered true.

In this program, there are two conditions separated by an OR operator. One is 'x = 3'; the other is '4'. '4' is not equal to 0 or missing, so it is evaluated as true. An OR expression is true if either one (or both) of the expressions is true. Because '4' is true, the logical expression is true regardless of the value of *x*.

Just to be sure this is clear, let's rewrite Program 11.4 correctly like this:

Program 11.5: A Corrected Version of Program 11.4

```
data Mystery;
   input x;
   if x = 3 or x = 4 then Match = 'Yes';
   else Match = 'No';
datalines;
3
4
9
.
-5
;
title "Listing of Data Set MYDERY";

proc print data=Mystery noobs;
run;
```

Seeing how easy it is to make this error when using multiple OR operators, you should be convinced that it is better to write the logical test as:

```
if x in (3,4) then Match = 'Yes';
```

Conclusion

Just about every program you write will need to use conditional logic. Using IF-THEN-ELSE statements and the Boolean operators NOT, AND, and OR allows you to evaluate complex conditions. Finally, remember to consider missing values when you write your logical expressions.

Problems

1. Using the Sashelp data set Fish, create a new, temporary SAS data set called Group_Fish that contains the variables Species, Weight, and Height. Using IF-THEN-ELSE logic, create a new variable called Group_Fish that places the fish into weight groups as follows:

   0 to 100=1, 101-200=2, 201-500=3, 501-1,000=4, and 1,001 and greater=5.

   Use PROC PRINT to list the first 10 observations from data set Group_Fish.

2. Run Program 11.2 in this chapter (listed below), adding statements to print one message if the variable Age is greater than 100 and another message if Age is missing. Include the Age value in the message for ages greater than 100.

```
data People;
   input @1  ID     $3.
         @4  Gender $1.
         @5   Age    3.
         @8  Height  2.
         @10 Weight  3.;
   if missing(Age) then Age_Group = .;
   else if Age le 20 then Age_Group = 1;
   else if Age le 40 then Age_Group = 2;
   else if Age le 60 then Age_Group = 3;
   else if Age le 80 then Age_Group = 4;
   else if Age ge 80 then Age_Group = 5;
datalines;
001M 5465220
002F10161 98
003M 1770201
004M 2569166
005F   64187
006F 3567135
;
```

3. Create a new, temporary SAS data set called High_BP containing subjects from the Sashelp.Heart data set who have systolic blood pressure greater than 250 or diastolic blood pressure greater than 180. The new data set should only have variables Diastolic, Systolic, and Status. Use PROC PRINT to list the contents of High_BP.

4. If A is true, B is false, and C is false, what do each of these expressions evaluate as (true or false)?

   a) A AND NOT B
   b) NOT A OR NOT C
   c) A AND NOT B AND C
   d) A AND (NOT B OR NOT C)

5. What's wrong with this program?

```
1. data Weights;
2.    input Wt;
3.    if Wt lt 100 then Wt_Group = 1;
```

```
4.    if Wt lt 200 then Wt_Group = 2;
5.    if Wt lt 300 then Wt_Group = 3;
6. datalines;
50
150
250
;
```

6. Starting with the Sashelp data set Retail, write a program to create a new data set (Sales_Status) with the following new variables:

   If Sales is greater than or equal to 300, set Bonus equal to 'Yes' and Level to 'High'. Otherwise, if Sales is not a missing value, set Bonus to 'No' and level to 'Low'. Use PROC PRINT to list the observations in this data set.

Chapter 12: Performing Iterative Processing: Looping

Introduction

Many programs use a process called *looping*. You can execute a series of statements a fixed number of times or you can set a condition that, when satisfied, will cause the iteration to start or stop and the rest of the program to continue. SAS iterative statements all start with the keyword DO and are referred to as *DO loops*.

Demonstrating a DO Group

Although this chapter deals mainly with DO loops, this is a good place to demonstrate a DO group. When you conduct a logical test, for example if `Gender = 'F'`, you may want to do several things. A DO group allows you to accomplish this.

In this example, you want to perform several actions when the value of LDL (low-density lipids, the "bad" cholesterol) is above 100.

```
if LDL gt 100 then do;
   Stoke_Risk = 'High';
   LDL_Group = 'Bad';
end;
```

Rather than repeating the IF statement two times, you create a DO group, starting with the keyword DO and ending with the keyword END. All the statements between DO and END execute when the value of LDL is greater than 100. It is standard programming practice to indent all the statements between the DO and END statements—it makes the program easier to read.

Describing a DO Loop

Let's start off with an example, similar to the one in Chapter 7, that converts temperatures in degrees Celsius to degrees Fahrenheit. Except this time you want to produce a table of temperatures from 0 degrees Celsius to 100 degrees Celsius.

Program 12.1: Creating a Table of Celsius and Fahrenheit Temperatures

```
data Convert_Temp;
   do Temp_C = 0 to 100;
      Temp_F = 1.8*Temp_C + 32;
      output;
   end;
run;

title "Listing of Data Set CONVERT_Temp";
proc print data=Convert_Temp noobs;
run;
```

The general form of a DO loop is:

```
do variable = lower-limit to upper-limit by increment;

   SAS statements;

end;
```

This is how it works. The SAS variable following the keyword DO is first given a value defined by the lower limit. Next, the statements between the DO and END execute. At the bottom of the loop (at the END statement), the value of the variable is incremented by the value of the increment. If an increment is not specified (that is, the BY variable is left out), the increment is assumed to be 1. If the value of the variable is greater than the upper limit, the loop is finished and the statement(s) following the END statement execute. If not, process control returns to the top of the loop and the statements in the DO loop execute again.

Let's follow the sequence of events in Program 12.1:

1. The variable Temp_C is given a value of 0.
2. The variable Temp_F is computed.
3. The OUTPUT statement writes an observation to the data set Convert_Temp.
4. At the END statement, the variable Temp_C is incremented by 1.
5. The program now loops back to the DO statement and checks to see if the value of Temp_C is greater than 100 (it is not).
6. Temp_C is set equal to 1 and a value is computed for Temp_F.
7. Another observation is written to the data set.
8. The loop continues until an observation with Temp_C equal to 100 is written out.
9. The variable Temp_C is incremented by 1 (now equal to 101).
10. Because 101 is greater than 100 (the upper limit), the DO loop terminates.
11. The program ends with the RUN statement.

> It should be noted that because an OUTPUT statement was used in this DATA step, the automatic output at the bottom of the DATA step does not occur. This prevents the last observation in the data set from being written twice.

A portion of the output is shown below:

Output from Program 12.1

Listing of Data Set CONVERT_Temp

Temp_C	Temp_F
0	32.0
1	33.8
2	35.6
3	37.4
4	39.2
5	41.0
6	42.8
7	44.6
8	46.4
9	48.2
10	50.0
11	51.8
12	53.6
13	

If you want to change the interval to 10 degrees Celsius, change the DO statement to read:

```
do Temp_C = 0 to 100 by 10;
```

The output would then look like this:

Output from Program 12.1 (with the Interval Set to 10)

Listing of Data Set CONVERT_Temp

Temp_C	Temp_F
0	32
10	50
20	68
30	86
40	104
50	122
60	140
70	158
80	176
90	194
100	212

Using a DO Loop to Graph an Equation

Suppose you want to create a graph of an equation and can't find the graphing calculator you used in high school. As an example, let's look at the following equation:

$$y = 2x^3 - x^2 + 3x$$

You can easily compute multiple values of *y* for a family of *x* values. Let's generate *y* values for *x* values ranging from -5 to 5 in intervals of .01:

Program 12.2: Graphing a Cubic Equation

```
data Cubic;
   do x = -5 to 5 by .01;
      y = 2*x**3 - x**2 + 3*x;
      output;
   end;
run;

title "Graph of Cubic Equation";
proc sgplot data=Cubic;
   series x=x y=y;
run;
```

Data set Cubic contains the *x* and *y* values computed by the DO loop. At this point, you could use the built-in tasks to produce a series plot or to write the statements using PROC SGPLOT to produce the plot. Here is the resulting plot:

Output from Program 12.2

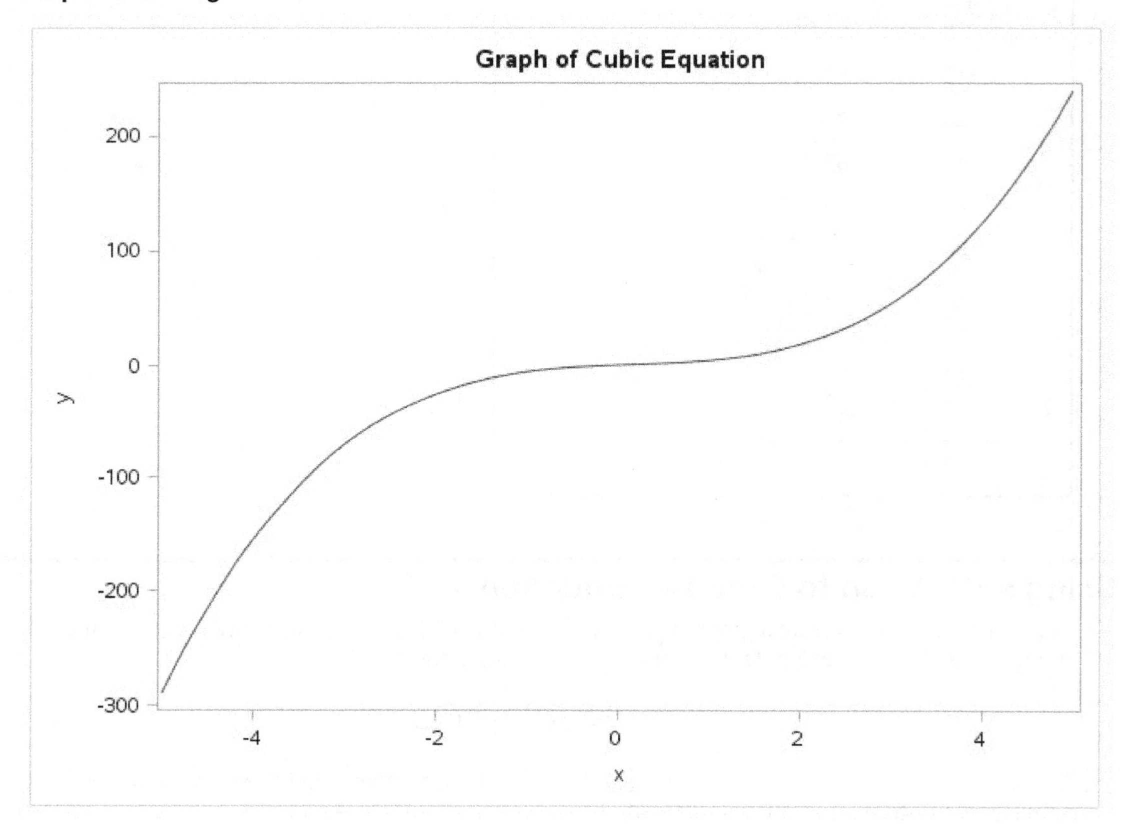

And there it is!

DO Loops with Character Values

Amazing as it may seem, you can write DO loops with character values. As an example, suppose you have heart rates for subjects given one of three drugs: Placebo, Drug A, and Drug B. Each line of data contains three heart rates, one for each of the three drug groups. Here is what the data looks like:

Data for Heart Rate Study

```
80 70 60
82 77 63
76 74 70
78 80 67
```

A program to read this data and assign values to a drug group variable is shown next:

Program 12.3: Using a DO Loop with Character Values

```
data HR_Study;
   do Drug_Group = 'Placebo','Drug A','Drug C';
      input Heart_Rate @;
      output;
   end;
datalines;
80 70 60
82 77 63
76 74 70
78 80 67
;
title "Listing of Data Set HR_Study";
proc print data=HR_Study noobs;
run;
```

Instead of a starting and ending value, you list each of the character values you want to assign to Drug_Group in quotes (single or double), separated by commas. The trailing @ sign on the INPUT statement prevents SAS from going to a new line each time the loop iterates and the INPUT statement is executed. The trailing @ sign ("hold the line") allows the program to read all three values from one line of data.

To help make this clear, here is a listing of data set HR_Study:

Listing of Data Set HR_Study

Drug_Group	Heart_Rate
Placebo	80
Drug A	70
Drug C	60
Placebo	82
Drug A	77
Drug C	63
Placebo	76
Drug A	74
Drug C	70
Placebo	78
Drug A	80
Drug C	67

Leaving a Loop Based on Conditions (DO WHILE and DO UNTIL Statements)

There are two statements that perform a logical test to decide when to leave a DO loop—DO WHILE and DO UNTIL. Here's how they work.

DO WHILE

The syntax for a DO WHILE loop is:

```
do while (expression);
   SAS Statements
End;
```

where *expression* is a SAS logical expression. This structure will loop from the DO statement to the END statement as long as the expression is true.

> In a DO WHILE statement, the *expression* is evaluated at the top of the loop; therefore, if the expression remains false, the SAS statements in the DO WHILE loop will never execute.

Here are some examples:

Program 12.4: Demonstrating a DO WHILE Loop

```
data Bank;
   Interest_Rate=.07;
   Amount = 1000;
   Goal = 2000;
   do while (Amount lt Goal);
      Year + 1;
      Amount = Amount + Interest_Rate*Amount;
      output;
   end;
   format Amount Goal dollar9.2;
run;

title "Listing of Data Set BANK";

proc print data=Bank noobs;
run;
```

You start with $1,000 in the bank. You want to see how many years it takes to reach (or exceed) your goal of $2,000, with interest set at 7%. Because this bank compounds interest each year, you multiply the interest rate by the cumulative amount each year. The loop stops when Amount is greater than or equal to $2,000. The statement `Year + 1;` is a SUM statement. It initializes Year at 0 and adds one each time the SUM statement executes. If this had been an ASSIGNMENT statement instead of a SUM statement, the variable Year would be set to a missing value at the top of the DATA step. Here is the listing:

Output from Program 12.4

Listing of Data Set BANK

Interest_Rate	Amount	Goal	Year
0.07	$1,070.00	$2,000.00	1
0.07	$1,144.90	$2,000.00	2
0.07	$1,225.04	$2,000.00	3
0.07	$1,310.80	$2,000.00	4
0.07	$1,402.55	$2,000.00	5
0.07	$1,500.73	$2,000.00	6
0.07	$1,605.78	$2,000.00	7
0.07	$1,718.19	$2,000.00	8
0.07	$1,838.46	$2,000.00	9
0.07	$1,967.15	$2,000.00	10
0.07	$2,104.85	$2,000.00	11

When the value of Amount is greater than Goal, the DO WHILE loop ends. (It's been a long time since interest rates were at 7%, but this author didn't want the listing to be too long.)

Combining an Iterative Loop with a WHILE Condition

> Caution! If you are not careful, DO WHILE and (especially) DO UNTIL loops may cause infinite loops. This can occur if the condition for a DO WHILE loop is always true or the condition for a DO UNTIL loop is never achieved.

You can combine a traditional DO loop with a WHILE condition. You may want to do this only while testing your code, or you may have reason to use both conditions. Here is an example:

Program 12.5: Combining an Iterative DO Loop with a WHILE Condition

```
data Bank;
   Interest_Rate=.07;
   Amount = 1000;
   Goal = 2000;
   do I = 1 to 20 while (Amount lt Goal);
      Year + 1;
      Amount = Amount + Interest_Rate*Amount;
      output;
   end;
   format Amount Goal dollar9.2;
   drop I;
run;

title "Listing of Data Set BANK";

proc print data=Bank noobs;
run;
```

Even if the WHILE condition remains true, this loop will stop after 20 iterations.

DO UNTIL

The companion to DO WHILE is DO UNTIL. There are some important differences. First of all, as you would expect, a DO UNTIL loop will continue to loop until the UNTIL condition is true. DO UNTIL loops are even more prone to infinite loops than DO WHILE loops.

> The UNTIL condition is evaluated at the bottom of the loop. Therefore, DO UNTIL loops will always execute at least one time.

The following program revisits the compound interest problem using a DO UNTIL. Here is the code:

Program 12.6: Demonstrating a DO UNTIL Loop

```
data Bank;
   Interest_Rate=.07;
   Amount = 1000;
   Goal = 2000;
   do until (Amount gt Goal);
      Year + 1;
      Amount = Amount + Interest_Rate*Amount;
      output;
   end;
   format Amount Goal dollar9.2;
run;

title "Listing of Data Set BANK";

proc print data=Bank noobs;
run;
```

The output from this program is identical to the output from Program 12.4.

Demonstrating That a DO UNTIL Loop Executes at Least Once

The short program below demonstrates that the statements in a DO UNTIL loop execute at least once. Here is the program:

Program 12.7: Demonstrating That a DO UNTIL Loop Will Always Execute Once

```
data At_Least_Once;
   x = 5;
   do until (x = 5);
      put "This line is inside the loop";
   end;
run;
```

When you run this program, the message "This line is inside the loop" prints once in the SAS log.

Combining an Iterative Loop with an UNTIL Condition

DO UNTIL loops are even more likely to cause infinite looping (a bad thing, especially if you are writing out to a file in the loop) than a DO WHILE loop. Therefore, you may want to "take out some insurance" by combining an iterative DO loop with an UNTIL condition. You can choose the number of iterations to be much larger than you need, but it is just there in case the UNTIL condition is never true. Here is an example:

Program 12.8: Combining an Iterative DO Loop with an UNTIL Condition

```
data Bank;
   Interest_Rate=.07;
   Amount = 1000;
   Goal = 2000;
   do I = 1 to 100 until (Amount gt Goal);
      Year + 1;
      Amount = Amount + Interest_Rate*Amount;
      output;
   end;
   format Amount Goal dollar9.2;
   drop I;
run;

title "Listing of Data Set BANK";

proc print data=Bank noobs;
run;
```

The output from this program is identical to the output from Program 12.6.

LEAVE and CONTINUE Statements

The two statements LEAVE and CONTINUE cause two different actions to take place within a loop. LEAVE, as the name suggests, jumps to the first line following the loop; CONTINUE skips the remaining statements in the loop, but it does return control back to the top of the loop. You can add a conditional LEAVE statement inside an ordinary DO loop and it can mimic a DO WHILE or DO UNTIL loop. Here is an example:

Program 12.9: Demonstrating a LEAVE Statement

```
data Bank;
   Interest_Rate=.07;
   Amount = 1000;
   Goal = 2000;
   do I = 1 to 20;
      Year + 1;
      Amount = Amount + Interest_Rate*Amount;
      output;
      if Amount gt Goal then leave;
   end;
   format Amount Goal dollar9.2;
   drop I;
run;

title "Listing of Data Set BANK";
```

```
proc print data=Bank noobs;
run;
```

Although this example produces the same output as Program 12.4, it lacks the esthetics of DO WHILE or DO UNTIL loops.

This next program demonstrates how a CONTINUE statement can be used. In this example, you want to compute compound interest rates, but you only want to see the dollar amounts once the amount is greater than $1,500. Here is the code:

Program 12.10: Demonstrating a CONTINUE Statement

```
data Bank;
   Interest_Rate=.07;
   Amount = 1000;
   Goal = 2000;
   do until (Amount gt Goal);
      Year + 1;
      Amount = Amount + Interest_Rate*Amount;
      if Amount lt 1500 then continue;
      output;
   end;
   format Amount Goal dollar9.2;
run;

title "Listing of Data Set BANK";
proc print data=bank noobs;
run;
```

As long as the Amount is less than $1,500, the OUTPUT statement is skipped, but the loop continues. Once the Amount is greater than or equal to $1,500, the OUTPUT statement executes.

Output from Program 12.10

Listing of Data Set BANK

Interest_Rate	Amount	Goal	Year
0.07	$1,500.73	$2,000.00	6
0.07	$1,605.78	$2,000.00	7
0.07	$1,718.19	$2,000.00	8
0.07	$1,838.46	$2,000.00	9
0.07	$1,967.15	$2,000.00	10
0.07	$2,104.85	$2,000.00	11

Conclusion

You have seen how to create a DO group and a DO loop as well as DO WHILE and DO UNTIL loops. It is important to remember that DO WHILE loops evaluate the logical expression at the top of the loop and DO UNTIL loops evaluate the logical expression at the bottom of the loop (thus, a DO UNTIL loop always executes at least once). Also, be careful to avoid infinite loops. They can possibly cause your computer to crash or a disk drive to fill to capacity (a really bad thing).

Problems

1. Write a DATA step to create a conversion table for pounds and kilograms. The table should have one column showing pounds from 0 to 100 in units of 10. The second column in the table should show the kilogram equivalents. Note: 1 Kg. = 2.2 Lbs.
2. Write a program to create a table with column 1 containing the integers from 1 to 10, column 2 the square root of column 1, and column 3 the square of column 1. Hint: To take a square root, you can use an exponent of .5 or a function called SQRT. To use the function, you would write something like: Root_x = sqrt(x);
3. You have data from three groups of subjects (Groups A, B, and C). The actual data looks like this:

Group	Score
A	10
B	11
C	12
A	20
B	21
C	22

However, the data was entered without the groups, like this:

Data for Group Study

10

11

12

20

21

22

Write a program to read the six numbers shown here (use DATALINES) to create a table with the variables Group and Score.

4. You have a variable called Money initialized at 100. Write a DO WHILE loop that compounds this amount by 3 percent each year and computes the amount of money plus interest for each year. Stop when the total amount exceeds 200.

   To help get you started, the beginning of the program should look like this:

Program for Problem Sets 1

```
data Interest;
   Money = 100;
   do while (put something here);
      Year + 1; *keep track of years;
      *compute new amount;
      output; *output an observation for each iteration
               of the loop;
   end;
run;
```

5. Solve Problem 4 using DO UNTIL instead of DO WHILE.
6. What is the value of Y when you run this program?

Program for Problem Sets 2

```
data Until;
   X = 5;
   Y = 10;
   do until (X eq 5);
      Y = 20;
   end;
run;
```

Chapter 13: Working with SAS Dates

Introduction

SAS can read and write dates in almost any format, such as 5/23/2015 or 23May2015. No matter how the date appears in the input file, SAS converts all dates to the number of days from January 1, 1960. Thus, January 1, 1960, is 0; January 2, 1960, is a 1; and so on. Dates before January 1, 1960, are converted to negative numbers. For example, December 31, 1959, is equal to -1. Almost all computer languages store dates as the number of days from a fixed date. SAS chose January 1, 1960—other languages use other dates. Although SAS dates are stored internally as numbers, SAS can display the date in any of the standard formats. For those readers interested in history, SAS does not compute dates before January 1, 1582. That is because Pope Gregory the VIII decreed that October 4, 1582, would be followed by October 15, 1582 (the beginning of the Gregorian calendar).

Reading Dates from Text Data

SAS uses informats to convert dates expressed in any one of the possible date formats into its internal representation of dates (the number of days from January 1, 1960). Here is an example:

You have a text file Date_Data.txt containing three dates, as follows:

File Date_Data.txt

```
123456789012345678901234567890 (Ruler, not part of the file)
05/23/2015 23May2015 23/05/2015
10/21/1950 21Oct1950 21/10/1950
5/7/2013   7Jul2013  7/5/2013
```

The first date, starting in column 1, is in the form *mm/dd/yyyy* (month, day, year). The second date, starting in column 12, is in the form *nnMonyyyy* (day of the month, month abbreviation, year). The third date, starting in column 22, is in the form *dd/mm/yyyy* (day, month, year). The following program reads each of these dates:

Program 13.1: Reading Dates in a Variety of Date Formats

```
data Read_Dates;
   infile '/folders/myfolders/Date_Data.txt' pad;
   input @1  Date1 mmddyy10.
         @12 Date2 date9.
         @22 Date3 ddmmyy10.;
run;

title "Listing of Data Set DATES";
proc print data=Read_Dates noobs;
run;
```

You use the appropriate informat to read each of the dates. The two informats mmddyy10. and ddmmyy10. are pretty obvious. By the way, the 10 at the end specifies that the program is reading 10 columns of data. If you had a date such as 5/6/2013 anywhere in the specified 10 columns, SAS would still read it correctly. The informat DATE9. reads dates in this form: day of the month (one or two digits), month abbreviation (not case-sensitive), and four-digit year.

Here is the listing:

Output from Program 13.1

Listing of Data Set DATES

Date1	Date2	Date3
20231	20231	20231
-3359	-3359	-3359
19485	19546	19485

This looks pretty strange. What you are seeing are the internal values for each of the dates—the number of days from January 1, 1960. To see these values as dates, you need to associate a SAS date format with each variable. Here is a list of some of the more popular date formats:

List of Popular Date Formats

How Date is Displayed	Date Format
10/21/1950	mmddyy10.
10-21-1950	mmddyyd10.
10 21 1950	mmddyyb10.
10:21:1950	mmddyyc10.
21/10/1950	ddmmyy10.
21Oct1950	DATE9.
Saturday, October 21, 1950	WEEKDATE.

Note: The formats that end in 10d, 10b, and 10c (rows 3-5) replace the default slash with dashes, blanks, or colons, respectively. You can add d's, b's, or c's to the ddmmyy formats as well.

You can use any date format you want to display any of these date variables. The program below is identical to Program 13.1, except that each of the dates is now formatted:

Program 13.2: Adding Formats to Display the Date Values

```
data Read_Dates;
   infile '/folders/myfolders/Date_Data.txt' pad;
   input @1  Date1 mmddyy10.
         @12 Date2 date9.
         @22 Date3 ddmmyy10.;
   format Date1 mmddyy10. Date2 Date3 date9.;
run;

title "Listing of Data Set DATES";
proc print data=Read_Dates noobs;
run;
```

The listing below shows the effect of these formats:

Output from Program 13.2

Listing of Data Set DATES

Date1	Date2	Date3
05/23/2015	23MAY2015	23MAY2015
10/21/1950	21OCT1950	21OCT1950
05/07/2013	07JUL2013	07MAY2013

The date values are now displayed properly. Because the month-day-year format is popular in the United States and the day-month-year format is popular in most of the world outside the United States, many companies that do global business prefer the DATE9. format (used for Date2 and Date3 in the listing).

Creating a SAS Date from Month, Day, and Year Values

If you have month, day, and year values as separate variables, you can use the MDY function to create a SAS date. This function takes three numeric arguments and, as the function name suggests, the three arguments provide values for month, day, and year.

Here is an example: You have a SAS data set with variables Month, Day, and Year. To compute a SAS date, use the following statement:

```
Date = MDY(Month, Day, Year);
```

The variable Date will be a SAS date (i.e., the number of days from January 1, 1960). You will most likely want to add a FORMAT statement to associate one of the date formats with the variable Date.

Describing a Date Constant

If you need to refer to a specific date in a DATA step, you could always enter that date as the number of days from January 1, 1960. However, computing that value is inconvenient and a program that referred to a date by its internal value would also be hard to read. The answer: The date constant.

Suppose you want to test a date to see if it is earlier than January 1, 2012, or later than December 31, 2014. Here's how to do it:

Program 13.3: Demonstrating a Date Constant

```
data _null_;
title "Checking for Out of Range Dates";
```

```
   input @1 Date mmddyy10.;
   file print;
   if Date lt '01Jan2012'd and not missing(Date) or
      Date gt '31Dec2014'd then put "Date " Date "is out of range";
   format Date mmddyy10.;
datalines;
10/13/2014
5/1/2011
1/1/2015
6/5/2014
1/1/2000
;
```

As demonstrated in this program, a date constant consists of a date in the form *nnMonyyyy*, where *nn* is the day of the month, *Mon* is the month abbreviation, and *yyyy* is the year. You place this value in single or double quotes and follow it (no spaces) with an uppercase or lowercase d. At compile time, SAS converts the date constants to SAS dates.

Before we get to the output, let's discuss a few features of the program. First of all, it uses the reserved data set name _NULL_. Giving a data set this special name has the effect of not creating any data set at all. That is, after the DATA step runs, there is no data set left behind. Using a _NULL_ statement saves the computer all the overhead of creating the data set and writing observations to it. Because of this, you can't use a PROC PRINT (or any other procedure) to list the contents of the data set, because there is no data set. The solution is to use a PUT statement. Following the keyword PUT, you can enter text (in quotes) and variable names. SAS will print out the quoted text and the value of the variables listed on the PUT statement. Finally, the statement FILE PRINT is an instruction to print the results to the RESULTS window. If you leave off this statement, SAS, by default, sends the results of the PUT statement to the SAS log.

As a reminder, a missing value (logically the most negative value you can have) is going to be less than any date. By using the MISSING function in this program, you are giving instructions not to print the out-of-range message if the date is a missing value.

Here is the output from Program 13.3:

Checking for Out of Range Dates

```
Date 05/01/2011 is out of range
Date 01/01/2015 is out of range
Date 01/01/2000 is out of range
```

The program works as advertised.

Extracting the Day of the Week, Day of the Month, and Year from a SAS Date

You can use the three functions, WEEKDAY, DAY, and YEAR, to compute the day of the week (a number from 1 to 7, with 1=Sunday), day of the month (a number from 1 to 31), and year from a SAS date. Let's look at an example.

You have a SAS data set with a variable called Date. You would like to generate bar charts showing frequencies for day of the week, day of the month, and year.

Note: To save room, this program only produces a bar chart for the day of the week. You can substitute the day of the month or year to obtain bar charts for these variables. The DATA step that creates the data set containing the Date variable is included so that you can try running the program yourself.

Program 13.4: Extracting the Day of the Week, Day of the Month, and Year from a SAS Date

```
data Extract;
   informat Date mmddyy10.;
   input Date @@; ①
   Day_of_Week = weekday(Date); ②
   Day_of_Month = day(Date);
   Year = year(Date);
   format Date mmddyyd10.;
datalines;
1/5/2000 2/8/2000 4/23/2000 4/12/2000 8/21/2000 8/21/2000 8/22/2000
12/12/2000 12/15/2000 12/18/2000
2/22/2001 2/1/2001 4/18/2001 4/18/2001 4/18/2001 9/17/2001
12/25/2001
12/22/2001 3/3/2001 3/6/2001 3/7/2001
;

title "Listing of the First Eight Observations from EXTRACT";
proc print data=Extract (obs=8); ③
run;

title "Frequencies for Day of the Week";
proc sgplot data=Extract; ④
   vbar Day_of_Week;
run;
```

① The two @ signs at the end of the INPUT statement prevent SAS from going to a new line when the DATA step iterates. It allows you to place data for more than one observation on a single line of data.

② The three functions, WEEKDAY, DAY, and YEAR, are used to extract the day of the week, the day of the month, and the year from the variable Date.

③ The data set option OBS= is an instruction to stop processing when you reach observation 8. It is a convenient way to list the first *n* observations in a data set.

④ You use PROC SGPLOT to generate a vertical bar chart (VBAR) for the variable Day_of_Week.

Here is the output from Program 13.4:

Output from Program 13.4

Listing of the First Eight Observations from EXTRACT

Obs	Date	Day_of_Week	Day_of_Month	Year
1	01-05-2000	4	5	2000
2	02-08-2000	3	8	2000
3	04-23-2000	1	23	2000
4	04-12-2000	4	12	2000
5	08-21-2000	2	21	2000
6	08-21-2000	2	21	2000
7	08-22-2000	3	22	2000
8	12-12-2000	3	12	2000

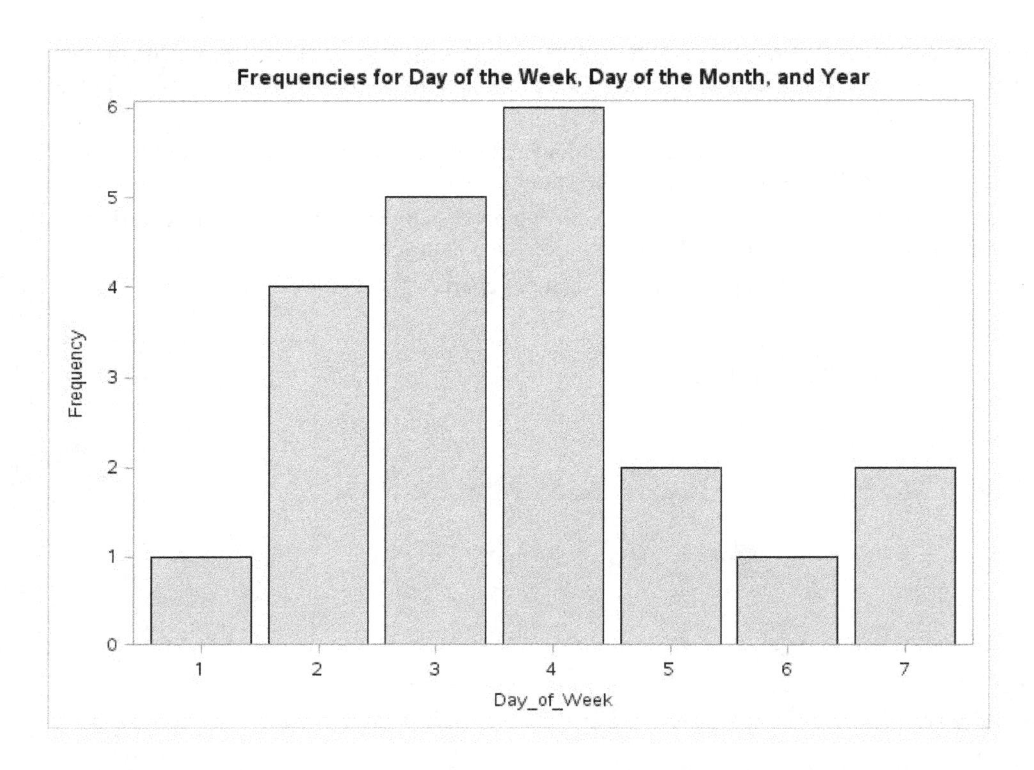

The first eight observations are printed. Notice the date format (it uses dashes because of the mmddyyd10. format) and the values for the other three variables.

The bar chart shows frequencies for the days of the week.

Adding a Format to the Bar Chart

To make the bar chart more readable, you can write a format that will substitute the day abbreviations (Mon, Tue, etc.) for the numbers 1–7. The program that follows does just that:

Program 13.5: Creating a Variable Representing Day of the Week Abbreviations

```
proc format;
   value dow 1='Sun' 2='Mon' 3='Wed' 4='Thu'
             5='Fri' 6='Sat' 7='Sun';
run;

data Extract;
   informat Date mmddyy10.;
   input Date @@;
   Day_of_Week = weekday(Date);
```

```
   Day_of_Month = day(Date);
   Year = year(Date);
   format Date mmddyyd10. Day_of_Week dow.;
datalines;
1/5/2000 2/8/2000 4/23/2000 4/12/2000 8/21/2000 8/21/2000 8/22/2000
12/12/2000 12/15/2000 12/18/2000
2/22/2001 2/1/2001 4/18/2001 4/18/2001 4/18/2001 9/17/2001
12/25/2001
12/22/2001 3/3/2001 3/6/2001 3/7/2001
;

title "Listing of the First Eight Observations from EXTRACT";
proc print data=Extract (obs=8);
run;

title "Frequencies for Day of the Week";
proc sgplot data=Extract;
   vbar Day_of_Week;
run;
```

The bar chart now looks like this:

Output from Program 13.5

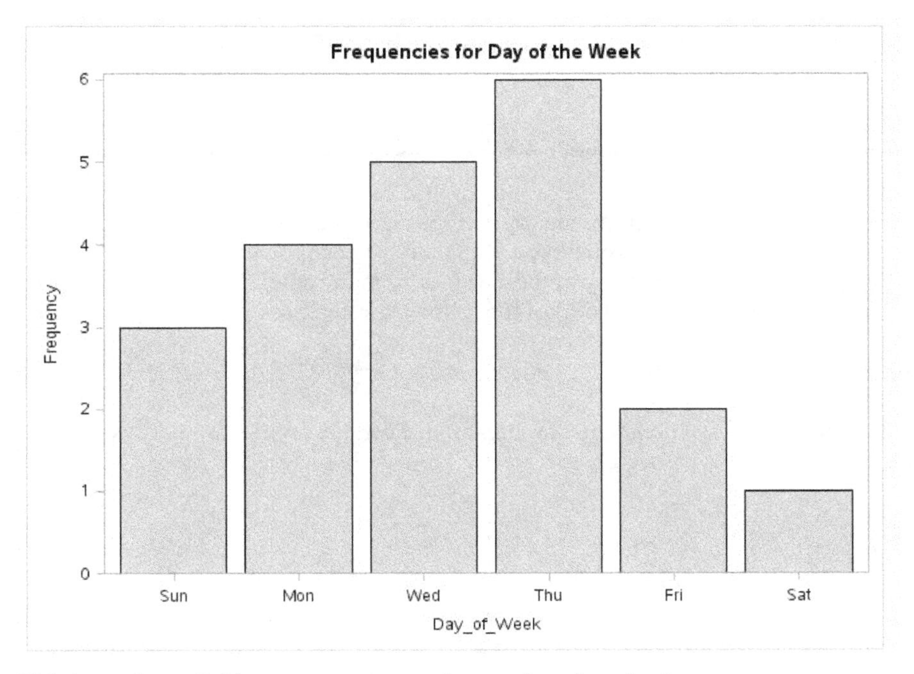

This is a substantial improvement over the previous bar chart.

Computing Age from Date of Birth: The YRDIF Function

Computing a person's age, given the date of birth, is a problem faced by just about every programmer. Luckily, SAS has a function called YRDIF (year difference) that computes the difference between two SAS dates in years. As an example, suppose you have a variable called DOB (date of birth) and you want to compute a person's age as of January 1, 2015. The calculation is:

```
Age = yrdif(DOB,'01Jan2015'd);
```

The two arguments to the YRDIF function are the first and last dates from which you want to compute the interval. In this example, the first date is a SAS date and the second date is a date constant. There are many applications where you need to compute age as of the last birthday. For example, you cannot vote even if your 18th birthday is the day after Election Day. You can use the INT (integer) function to extract the integer part of a number. The expression to compute a person's age on January 1, 2015, as of his or her last birthday is:

```
Age_Last = int(yrdif(DOB,'01Jan2015'd));
```

Notice that it is OK to have a SAS function as an argument to another SAS function. Just be careful with parentheses when you do this. You may want to draw the line at two functions in a single SAS statement. More than this can be a bit confusing to read.

If you need to round the result of the YRDIF function, use the ROUND function in place of the INT function.

Conclusion

SAS stores dates as the number of days from January 1, 1960. SAS has informats to read and interpret dates in almost any form. Once you have a SAS date, you can use the WEEKDAY, DAY, MONTH, and YEAR functions to extract any of these values from the date. To learn more about SAS date functions, please refer to one (or both) of the following references:

Cody, Ron. 2010. *SAS Functions by Example, Second Edition.* Cary, NC: SAS Institute Inc.

Morgan, Derek. 2014. *The Essential Guide to SAS Dates and Times, Second Edition.* Cary, NC: SAS Institute Inc.

Problems

1. You have a raw data file with the following data:

```
12345678901234567890 Ruler - not part of the data
10/21/2015 12Jun2015
12/25/2015  9Apr2014
```

   The date starting in column 1 is in the form month/day/year. The date starting in column 12 is in the form 2-digit day of the month, 3-character month abbreviation, and 4-digit year.

   Write an INPUT statement to read these two lines of data (you can use DATALINES). Call the first date Date1 and the second date Date2. Format both dates with the mmddyy10. format.

2. Run the program below to create a data set called Date_Test:

   Program for Problem Sets 1

```
data Date_Test;
   input Month Day Year;
datalines;
10 21 1988
3 4 2015
1 1 1960
;
```

   Modify this program so that you have a variable called Date that is a SAS date. Format this date using the DATE9. format.

3. The Sashelp data set Retail contains the variables Month, Day, and Year. Create a new, temporary data set called Dates that has these three variables plus one other called SAS_Date that is a true SAS date. Format this variable using the mmddyy10. format and list the first five observations from this data set. The Retail data set contains other variables besides Month, Day, and Year. You do not want any of these other variables in your Dates data set.
4. Using the data from Problem 1, compute three new variables, Month, Day (day of the week), and Year based on the date starting in column 1. Compute frequencies for these variables.
5. Run the following program to create a data set called Study:

   Program for Problem Sets 2

```
data Study;
   call streaminit(13579);
   do Subj = 1 to 10;
      Date = '01Jan2015'd + int(rand('uniform')*300);
      output;
   end;
   format Dates date9.;
run;
```

Write a DATA step that will print out all dates in the Study data set that are before January 1, 2015 or after July 4, 2015. You can either use a DATA _NULL_ DATA step with a PUT statement or create a data set of out-of-range dates and use PROC PRINT to print it.

6. Run the program in Problem 5 except change the line that computes dates to:

```
Dates = '01Jan1950'd + int(rand('uniform')*15000);
```

Assuming these dates represent the date of birth, compute the age of each subject as of January 1, 2015. Print out the subject numbers (variable Subj) and ages for each person. Add a FORMAT statement to your PROC PRINT to assign the format 4.1 to the variable Age (which will list each age, rounded to a tenth of a year).

Chapter 14: Subsetting and Combining SAS Data Sets

Introduction

This chapter discusses ways to subset (filter) a SAS data set and to combine data from several SAS data sets. For those readers who understand SQL (structured query language), you should know that SAS supports SQL in PROC SQL. Because SQL is covered elsewhere in many texts, this chapter does not include a discussion of PROC SQL. The methods described in this chapter, SET, MERGE, and UPDATE, are DATA step statements that provide an alternative to PROC SQL.

Subsetting (Filtering) Data in a SAS Data Set

As you saw in the first half of this book, you can subset or filter a SAS data set using interactive point-and-click operations provided by SAS Studio. This section shows you how to subset a SAS data set using DATA step programming.

To demonstrate how to create a subset starting with observations from an existing SAS data set, let's use one of the built-in data sets that comes with SAS Studio. There is a library called Sashelp that ships with SAS Studio. To see a list of data sets in this library, perform the following steps: First, click on the **Libraries** tab in the SAS Studio window.

Figure 14.1: Select the Libraries Tab

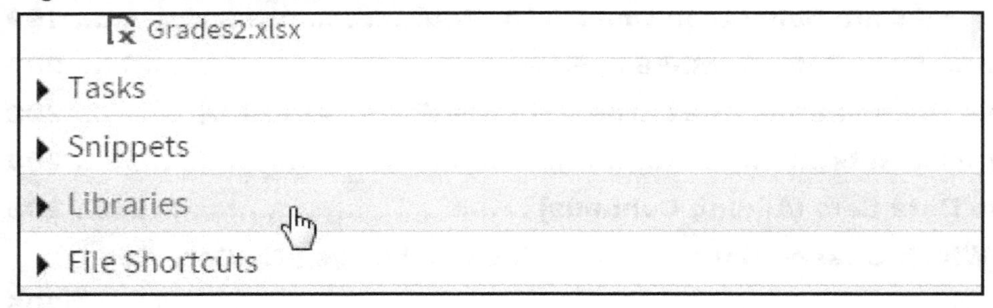

After you click on the **Libraries** tab, select **My Libraries** and then **SASHELP**. You will then see a list of the data sets in the Sashelp library.

Figure 14.2: Opening the Libraries Tab

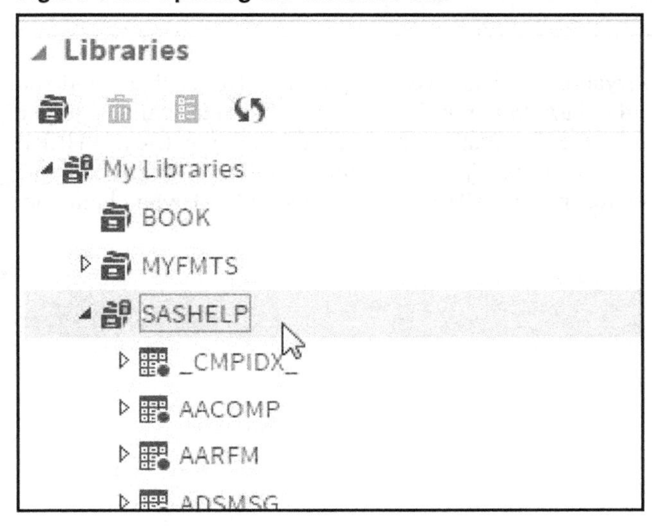

We are going to select the Retail data set. Scroll down the list of data sets and select **RETAIL**.

Figure 14.3: Opening the Retail Data Set

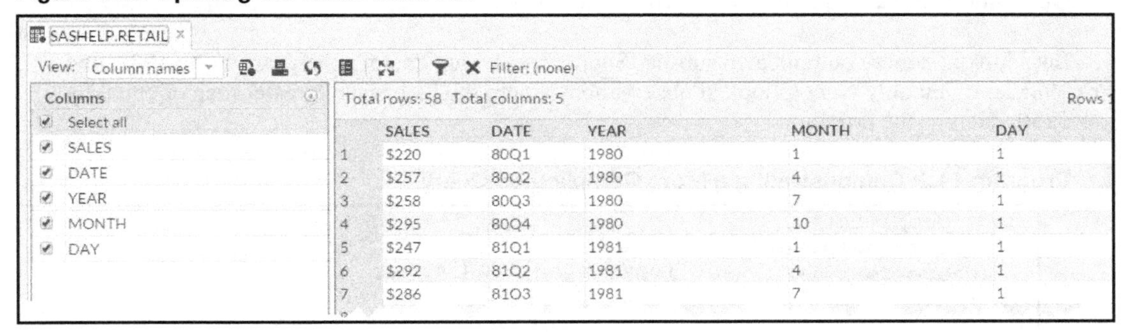

	SALES	DATE	YEAR	MONTH	DAY
1	$220	80Q1	1980	1	1
2	$257	80Q2	1980	4	1
3	$258	80Q3	1980	7	1
4	$295	80Q4	1980	10	1
5	$247	81Q1	1981	1	1
6	$292	81Q2	1981	4	1
7	$286	81Q3	1981	7	1

You see the variables in this data set along with a listing of the data set.

Suppose your goal is to concentrate on data from the year 1980. You proceed as follows:

Program 14.1: Using a WHERE Statement to Subset a SAS Data Set

```
data Year1980;
   set sashelp.retail;
   where Year = 1980;
run;

title "Listing of Data Set Year1980";
```

```
proc print data=Year1980 noobs;
run;
```

You use a SET statement to read the observations in data set Sashelp.Retail. Notice that you do not need to submit a LIBNAME statement—the Sashelp library is automatically made available to you every time you open SAS Studio. To subset the observations in data set Retail, you use a WHERE statement. Following the keyword WHERE, you specify the subsetting condition. In this example, you specify that you only want to see data for the year 1980. Using PROC PRINT, you obtain the following listing:

Output from Program 14.1

Listing of Data Set Year1980

SALES	DATE	YEAR	MONTH	DAY
$220	80Q1	1980	1	1
$257	80Q2	1980	4	1
$258	80Q3	1980	7	1
$295	80Q4	1980	10	1

Only four observations met your selection criteria.

Let's look at a more complicated subset. Suppose you want data for the years 1980, 1982, and 1985, and you only want to look at observations where the Sales were greater than or equal to $250. Here is the program:

Program 14.2: Demonstrating a More Complicated Query

```
data Complicated;
   set sashelp.retail;
   where Year in (1980, 1983, 1985) and Sales ge 250;
run;

title "Listing of Data Set Complicated";
proc print data=Complicated noobs;
run;
```

This WHERE statement is more complicated. You use an IN operator to select any observations where Year is equal to 1980, 1983, or 1985. Because you also want to restrict observations where Sales are greater than $250, you use the AND Boolean operator to add this condition. Here is the listing:

Output from Program 14.2

Listing of Data Set Complicated

SALES	DATE	YEAR	MONTH	DAY
$257	80Q2	1980	4	1
$258	80Q3	1980	7	1
$295	80Q4	1980	10	1
$299	83Q1	1983	1	1
$351	83Q2	1983	4	1
$359	83Q3	1983	7	1
$384	83Q4	1983	10	1
$337	85Q1	1985	1	1
$399	85Q2	1985	4	1
$412	85Q3	1985	7	1
$448	85Q4	1985	10	1

Describing a WHERE= Data Set Option

An alternative to a WHERE statement is a WHERE= data set option. This data set option is one of many possible data set options available to you. You place all of the data set options you wish to use in a set of parentheses following the data set name. For example, you can rewrite Program 14.1 using a WHERE= data set option like this:

Program 14.3: Rewriting Program 14.1 Using a WHERE= Data Set Option

```
data Year1980;
   set sashelp.retail (where=(Year = 1980));
run;
```

You need to be careful with parentheses. The outermost set of parentheses encloses any data set options you select—the inner set of parentheses encloses the WHERE condition. You use an equal sign following the keyword WHERE when you are subsetting using a data set option. You do not use an equal sign when you are writing a WHERE statement. The resulting data set (Year1980) is identical to the one created in Program 14.1.

By the way, you can use a WHERE statement or a WHERE= data set option in a SAS procedure. If you only want a listing of the 1980s data from the Sashelp.Retail data set, you can use the following statements:

Program 14.4: Using a WHERE= Data Set Option in a SAS Procedure

```
proc print data=sashelp.Retail (where=(Year = 1980));
run;
```

Describing a Subsetting IF Statement

Suppose you have some raw data consisting of ID, gender, age, height, and weight. Earlier in this book, you developed a simple DATA step to read this collection of data. Suppose you want to create a SAS data set from the raw data, but you only want to see the female subjects. You can use a subsetting IF statement as follows:

Program 14.5: Demonstrating the Subsetting IF Statement

```
data Females;
   input @1  ID     $3.
         @4  Gender $1.
         @5   Age    3.
         @8  Height  2.
         @10 Weight  3.;
   if Gender = 'F';
datalines;
001M 5465220
002F10161 98
003M 1770201
004M 2569166
005F   64187
006F 3567135
;
title "Listing of Data Set FEMALES";
proc print data=Females;
   id ID;
run;
```

Notice that the IF statement does not have a THEN clause. This special IF statement, known as a *subsetting IF*, works like this: If the statement is true, the program continues; if the statement is false, the program returns to the top of the DATA step to read another line of data. Because an implicit output is performed at the bottom of the DATA step, only females will be in the resulting data set. Here is the listing:

Output from Program 14.5

Listing of Data Set FEMALES

ID	Gender	Age	Height	Weight
002	F	101	61	98
005	F	.	64	187
006	F	35	67	135

A More Efficient Way to Subset Data When Reading Raw Data

Being a compulsive programmer, this author can't leave this section without showing you a more efficient way to write Program 14.5. Why read all of the data for males when you are not going to keep it? It is better to read just the single byte of data for Gender and only if it is equal to 'F' read the remaining data. Here is the program—the explanation will follow:

Program 14.6: Demonstrating a Trailing @

```
data Females;
   input @4  Gender $1. @;
   if Gender = 'F' then
      input @5   Age    3.
            @8  Height  2.
            @10 Weight  3.;
   else delete;
datalines;
001M 5465220
002F10161 98
003M 1770201
004M 2569166
005F   64187
006F 3567135
;
```

You may ask, "What is the @ sign doing at the end of the first INPUT statement?" Here's the explanation: If you have more than one INPUT statement in a DATA step, SAS will go to the next line of data each time it encounters another INPUT statement. The @ sign at the end of the line is called a *trailing @* and it tells SAS to "hold the line" and do not go to the next line of data when you encounter the next INPUT statement. Without the trailing @, the program would read the value of Gender from one line and input the other values from the next line.

The trailing @ on the first INPUT statement allows you to test the value of Gender and if it is equal to 'F' to read the values of Age, Height, and Weight from the **same line of data**. If Gender is not equal to 'F', the program executes the DELETE statement. This causes a return to the top of the DATA step without having to read the other values on the input record. Keep the trailing @ in mind whenever you need to read a portion of your data to decide how to read other values from the same line.

Creating Several Data Subsets in One DATA Step

If you need to create several subsets from a single SAS data set, you can reduce processing time by creating multiple SAS data sets in one DATA step. Doing this reduces processing time because you read through the data just once instead of multiple times. This is especially important when you are dealing with very large files. As an example, suppose you want to create three data sets from the Sashelp.Retail data set, each for a separate year. Here's how to do it:

Program 14.7: Creating Several SAS Data Sets in One DATA Step

```
data Year1980 Year1981 Year1982;
   set sashelp.Retail;
   if Year = 1980 then output Year1980;
   else if Year = 1981 then output Year1981;
   else if Year = 1982 then output Year1982;
run;
```

You name each of the data sets you want to create on the DATA statement. Next, you test the value of Year and use an OUTPUT statement, outputting observations to the appropriate data set. It is important to name the data set on the OUTPUT statement. If you use an OUTPUT statement without naming which data set you are writing to, the program will output to each of the data sets named on the DATA statement. One additional and important feature of this program is that when you include an OUTPUT statement in a DATA step, SAS does not perform an automatic output at the bottom of the DATA step.

When you run Program 14.7, the three data sets, Year1980, Year1981, and Year1982, are created.

Combining SAS Data Sets (Combining Rows)

One way to combine two or more data sets is to "stack them up" one on top of the other (also referred to as *concatenation*). For example, you may have data sets for each of the four quarters of the year and want to put them together to create a data set with all the data for the year. The simple example that follows shows how to do this using a SAS DATA step.

For this example, you have two data sets (cleverly called One and Two). They are shown below:

Figure 14.4: Data Sets One and Two

Listing of Data Set ONE

ID	DOB	Gender	Height	Weight
001	10/21/1950	M	68	160
002	11/11/1981	F	62	120
003	01/05/1983	M	72	220

Listing of Data Set TWO

ID	DOB	Gender	Height	Weight
004	05/13/1978	M	70	190
005	08/23/1988	F	59	98

You use a SET statement to combine these two data sets, one after the other, like this:

Program 14.8: Using a SET Statement to Combine Two SAS Data Sets

```
data Both;
   set One Two;
run;
```

You list each of the data sets you want to combine on the SET statement. The program first reads all the observations from data set One. When it reaches the end of the file, it switches to data set Two and continues to read observations from data set Two until it reaches the end of that file. The result is all the observations from the two data sets.

Here is a listing of data set Both:

Listing of Data Set Both

Listing of Data Set BOTH

ID	DOB	Gender	Height	Weight
001	10/21/1950	M	68	160
002	11/11/1981	F	62	120
003	01/05/1983	M	72	220
004	05/13/1978	M	70	190
005	08/23/1988	F	59	98

In case you want to run this program as is or with modifications, the program to create data sets One and Two is listed here:

Program 14.9: Program to Create Data Sets One and Two

```
data One;
   informat ID $3. DOB mmddyy10. Gender $1.;
   input ID DOB Gender Height Weight;
   format DOB mmddyy10.;
datalines;
001 10/21/1950 M 68 160
002 11/11/1981 F 62 120
003 1/5/1983 M 72 220
;

data Two;
   informat ID $3. DOB mmddyy10. Gender $1.;
   input ID DOB Gender Height Weight;
   format DOB mmddyy10.;
datalines;
004 5/13/1978 M 70 190
005 8/23/1988 F 59 98
;
```

Adding a Few Observations to a Large Data Set (PROC APPEND)

If your goal is to add observations (typically from a relatively small data set) to a large data set, you can, of course, use a SET statement and list the names of the two data sets, like this:

Program 14.10: Using a SET Statement to Solve the Problem

```
Data Both;
   Set Big Small;
run;
```

When you run this program, SAS will first read all the observations from data set Big and, when it reaches the end of the file, it will then read all the observations from data set Small.

If this is something that you need to do often, you may be concerned about efficiency, especially if data set Big is really big (millions or tens of millions of observations). A much more efficient method is to use PROC APPEND. A program that accomplishes the same goal as Program 14.10 (almost) is demonstrated in Program 14.11.

Program 14.11: Using PROC APPEND to Add Observations from One Data Set to Another

```
proc append base=Big Data=Small;
run;
```

You name the first data set on the BASE= procedure option and the data set to be added on the DATA= procedure option. This program **does not need to read any observations from data set Big**—it strips the end-of-file marker from the end of data set Big and then appends the new observations. If data set Big is large, this is a huge improvement in efficiency over using a SET statement. As with most things that look really quick and easy, there is a catch: In this program, you are replacing data set Big with the contents of Big and Small. If you have errors or bad data in data set Small, it is not easy to undo the process. It would be a good idea to have a backup copy of Big somewhere.

This method should only be used when you are confident that: One, the data you are adding does not contain errors; and two, data set Small has all the same variables and attributes (especially character variable lengths) as data set Big. The reason for this is that the data descriptor (containing all the information such as variable types and lengths) is taken from data set Big. If, for example, you had a character variable in data set Small that had a longer length than a variable of the same name in data set Big, that variable would be truncated. As a matter of fact, PROC APPEND would not even run in this situation unless you added an option called FORCE. The opinion of this author is that if you need to use the FORCE option, you had better know exactly what you are doing. One final point: If you have variables that have different names in the two data sets but are otherwise equivalent, you can use a RENAME= data set option to rename variables in one data set to match the variable names in the other data set.

Interleaving Data Sets

If you have two or more data sets that are so large that, when put together, it would be difficult or impossible to sort the result, you can sort each of the data sets first and then interleave them. That is, you can add observations to the resulting data set from each of the constituent data sets so that the result is in sorted order. All that is necessary to accomplish this is to follow the SET statement with a BY statement, listing the variables that define the sorted order. Here is an example:

Program 14.12: Demonstrating How to Interleave Two or More Data Sets

```
*Note: data sets One, Two, and Three are sorted by ID;
data combined;
   set One Two Three;
   by ID;
run;
```

The resulting data set (Combined) will be in ID order.

Merging Two Data Sets (Adding Columns)

By *merging*, we usually mean adding extra columns (variables) to a SAS data set based on one or more matching variables (such as ID or Name) from another data set. In SQL terms, you would be doing a *join* (inner, left, right, or, full). Imagine you are collecting clinical data on some patients. In one data set, you have ID, gender, and date of birth—in another data set, you have measurements such as weight, heart rate, and blood pressure. When it is time to analyze your data, you want to combine data from these two data sets. The program below creates two data sets, Patients and Visits.

Program 14.13: Creating the Patients and Visits Data Sets

```
data Patients;
   informat ID $4. Gender $1. DOB mmddyy10.;
   input ID Gender DOB;
   format DOB date9.;
datalines;
0001 M 10/10/1980
0023 F 1/2/1977
1243 M 6/17/2000
0002 M 8/23/1981
4535 F 2/25/1967
;

data Visits;
   informat ID $4. Visit_Date mmddyy10.;
   input ID Visit_Date Weight HR SBP DBP;
   format Visit_Date date9.;
datalines;
0023 2/10/2015 122 76 122 78
```

```
4535 10/21/2014 155 78 138 88
0001 11/11/2014 210 68 118 78
;
```

A listing of these two data sets is shown next:

Listing of Data Sets Patients and Visits

Listing of Data Set PATIENTS

ID	Gender	DOB
0001	M	10OCT1980
0023	F	02JAN1977
1243	M	17JUN2000
0002	M	23AUG1981
4535	F	25FEB1967

Listing of Data Set VISITS

ID	Visit_Date	Weight	HR	SBP	DBP
0023	10FEB2015	122	76	122	78
4535	21OCT2014	155	78	138	88
0001	11NOV2014	210	68	118	78

In order to combine (merge) these two data sets, you must first sort each of them by the variable (or variables) that you plan to use to match the two data sets. Here is the program, followed by an explanation:

Program 14.14: Merging Two SAS Data Sets

```
proc sort data=Patients;
   by ID;
run;

proc sort data=Visits;
   by ID;
run;

data Merged;
   merge Patients Visits;
   by ID;
run;

title "Listing of Data Set MERGED";
```

```
proc print data=Merged;
   id ID;
run;
```

You use PROC SORT to sort both data sets by ID. Next, you use a MERGE statement to combine the patient and visit data and a BY statement to instruct the program to use ID as the matching variable. Here is the resulting data set:

Output from Program 14.14

Listing of Data Set MERGED

ID	Gender	DOB	Visit_Date	Weight	HR	SBP	DBP
0001	M	10OCT1980	11NOV2014	210	68	118	78
0002	M	23AUG1981	.	.	.	.	.
0023	F	02JAN1977	10FEB2015	122	76	122	78
1243	M	17JUN2000	.	.	.	.	.
4535	F	25FEB1967	21OCT2014	155	78	138	88

What you see here is all the observations from the Patients data set matched with all the observations in the Visits data set (known as a *full join* in SQL), even if there is no corresponding visit for a particular ID.

This is probably not what you want. In this example, you most likely want to see a listing of only those patients who were in the Visits data set. Luckily, SAS has a way to control which observations will be included in the merged data set. Enter the IN= data set option.

Controlling Which Observations Are Included in a Merge (IN= Data Set Option)

You have previously seen a WHERE= data set option. You can now add to your collection of data set options by learning about the IN= data set option. Here's how it works:

When you are merging two data sets, you may find a matching ID in both data sets or you may only find an ID in one but not the other data set. To determine if a data set is making a contribution on a particular merge, you add an IN= *variable_name* data set option following each of the two data sets on the MERGE statement. Let's examine the value of these variables (called *contributor variables* by some SAS programmers) in the following program:

Program 14.15: Demonstrating the IN= Data Set Option

```
*Note: both data set previously sorted by ID;
data Merged;
   merge Patients(in=In_Patients) Visits(In=In_Visits);
   by ID;
   put ID= In_Patients= In_Visits=;
run;
```

The variable name following the IN= data set option can be any valid SAS name. However, this author likes to use names that help you remember which variable goes with which data set. Hence, the names In_Patients and In_Visits.

These two variables are temporary variables. That is, they are not included in the output data set—they are available only during the DATA step to help you determine how to perform the merge. A PUT statement was added to Program 14.15 so that you can examine these two variables. Here is the relevant section of the SAS log:

Partial Listing of the SAS Log from Program 14.15

```
39          data Merged;
40             merge Patients(in=In_Patients) Visits(In=In_Visits);
41             by ID;
42             put ID= In_Patients= In_Visits=;
43          run;
ID=0001 In_Patients=1 In_Visits=1
ID=0002 In_Patients=1 In_Visits=0
ID=0023 In_Patients=1 In_Visits=1
ID=1243 In_Patients=1 In_Visits=0
ID=4535 In_Patients=1 In_Visits=1
```

In each observation, the IN= variable is a 1 when that data set is contributing to the merge and 0 otherwise. Patient 0001 is in both data sets, so the value of the two IN= variables is 1. Patient 0002 is in the Patients data set but not the Visits data set, so In_Patients is a 1 and In_Visits is a 0 for this observation.

You can use the IN= variables to control which observations you want to include in the output data set. For example, to include only observations where there is a contribution from both data sets (referred to as an *inner join* is SQL), you would use:

```
if In_Patients=1 and In_Visits=1;
```

This can also be written as:

```
if In_Patients and In_Visits;
```

You do not need to include the equals 1 because SAS understands that these variables are either 1 or 0 (interpreted as true or false).

In this example, every patient in the Visits data set was matched to a record in the Patients data set. In the real world, this would not always be the case. The next program creates a data set containing patients in the Visits data set who are matched with the data in the Patients data set.

To make the program more general, statements were added to print an error message for any patient in the Visits data set who is missing from the Patients data set. It is good programming to expect the unexpected! Here is the program:

Program 14.16: Using the IN= Data Set Option to Control the Merge Operation

```
*Note: Both data sets already sorted by ID;
title "Listing of Data Set Only_Visit_Patients";
data Only_Visit_Patients;
   file print;
   merge Patients(in=In_Patients) Visits(in=In_visits);
   by ID;
   if In_Visits then output;
   if In_Visits and not In_Patients then
      put "Patient " ID "not found in the Patients data set.";
run;

title "Listing of Data Set ONLY_VISIT_PATIENTS";
proc print data=Only_Visit_Patients;
   id ID;
run;
```

You test the value of the variable In_Visits, and only output observations where In_Visits is true. If there was an ID in the Visits data set that did not have an observation in the Patients data set, an error message would be written. By default, PUT statements write data to the SAS log. By including the statement FILE PRINT, the output from this PUT statement will be sent to the RESULTS window. Here is the output:

Output from Program 14.16

Listing of Data Set ONLY_VISIT_PATIENTS

ID	Gender	DOB	Visit_Date	Weight	HR	SBP	DBP
0001	M	10OCT1980	11NOV2014	210	68	118	78
0023	F	02JAN1977	10FEB2015	122	76	122	78
4535	F	25FEB1967	21OCT2014	155	78	138	88

Performing a One-to-Many or Many-to-One Merge

If want to merge two data sets where one data set has more than one observation for your choice of BY variable(s) and the other data set has exactly one observation for the same BY variable(s), you can still perform a merge. For this example, you want to merge patient data (ID, Gender, and DOB) with visit data (ID, date of visit, HR, SBP, and DBP). To demonstrate this merge, we will use the Patients data set (see Program 14.13) and a new data set, Many_Visits, which contains several visits for each patient. The program to create the Many_Visits data set is shown in Program 14.17

Program 14.17: Creating Another Data Set to Demonstrate a One-to-Many Merge

```
data Many_Visits;
   informat ID $4. Visit_Date mmddyy10.;
   input ID Visit_Date HR SBP DBP;
   format Visit_Date date9.;
;
datalines;
0023 2/10/2015 122 76 122 78
0023 3/10/2015 120 74 120 76
4535 10/21/2014 155 78 138 88
0001 11/11/2014 210 68 118 78
0001 12/20/2014 210 68 120 82
0001 1/5/2015 212 70 210 80
;
```

For reference, both data sets (Patients and Many_Visits) are displayed below:

Listing of Data Sets Patients and Many_Visits

Listing of Data Set PATIENTS

ID	Gender	DOB
0001	M	10OCT1980
0023	F	02JAN1977
1243	M	17JUN2000
0002	M	23AUG1981
4535	F	25FEB1967

Listing of Data Set MANY_VISITS

ID	Visit_Date	HR	SBP	DBP
0023	10FEB2015	122	76	122
0023	10MAR2015	120	74	120
4535	21OCT2014	155	78	138
0001	11NOV2014	210	68	118
0001	20DEC2014	210	68	120
0001	05JAN2015	212	70	210

You can now sort and merge the two data sets like this:

Program 14.18: Performing a One-to-Many Merge

```
proc sort data=Patients;
   by ID;
```

```
run;

proc sort data=Many_Visits;
   by ID;
run;

data One_to_Many;
   merge Patients Many_Visits;
   by ID;
run;

title "Listing of data set ONE_TO_MANY";
proc print data=One_to_Many;
   id ID;
run;
```

When you run this program, the variables Gender and DOB from the Patients data set are combined with matching observations from the Many_Visits data set (based on ID) to create the data set One_To_Many, as shown in the listing below:

Output from Program 14.18

Listing of data set ONE_TO_MANY

ID	Gender	DOB	Visit_Date	HR	SBP	DBP
0001	M	10OCT1980	11NOV2014	210	68	118
0001	M	10OCT1980	20DEC2014	210	68	120
0001	M	10OCT1980	05JAN2015	212	70	210
0002	M	23AUG1981	.	.	.	.
0023	F	02JAN1977	10FEB2015	122	76	122
0023	F	02JAN1977	10MAR2015	120	74	120
1243	M	17JUN2000	.	.	.	.
4535	F	25FEB1967	21OCT2014	155	78	138

In this example, you could reverse the order of the two data sets (performing a many-to-one merge) and the result would be identical to the one obtained here.

Caution: Do not attempt to merge two data sets where there are multiple observations for each BY variable in both data sets and the number of multiples is not the same in both files.

Merging Two Data Sets with Different BY Variable Names

You may find yourself needing to merge two data sets, but the BY variable has a different variable name in each of the two files (this is corollary number 17 of Murphy's Law). The solution to this problem is actually very simple. Let's redo Program 14.14, but make a slight change to the data set Visits. The new data set (Visits_2) has all the same values of data set Visits except that the ID variable is named Pt.

Here is a listing of data set Visits_2:

Listing of Data Set Visits_2

Pt	Visit_Date	Weight	HR	SBP	DBP
0023	10FEB2015	122	76	122	78
4535	21OCT2014	155	78	138	88
0001	11NOV2014	210	68	118	78

Before we continue, here is the program that was used to create the data set Visits_2 (in case you want to try this yourself):

Program 14.19: Program to Create Data Set Visits_2

```
data Visits_2;
   informat Pt $4. Visit_Date mmddyy10.;
   input Pt Visit_Date Weight HR SBP DBP;
   format Visit_Date date9.;
datalines;
0023 2/10/2015 122 76 122 78
4535 10/21/2014 155 78 138 88
0001 11/11/2014 210 68 118 78
;
```

In order to merge the two files Patients and Visits_2, you have to rename one of the variables (either ID or Pt) so that both BY variables have the same name. You do this with a RENAME= data set option. In this example, you are going to rename the variable Pt in the Visits_2 data set to ID.

Program 14.20: Using a RENAME= Data Set Option to Rename the Variable Pt to ID

```
proc sort data=Patients;
   by ID;
run;

proc sort data=Visits_2;
   by Pt;
run;
```

```
data Merged;
   merge Patients Visits_2(rename=(Pt = ID));
by ID;
run;

title "Listing of Data Set MERGED";
proc print data=Merged;
   id ID;
run;
```

This is another opportunity to get confused with the parentheses. Remember, the outer-most set holds all of the data set options—the inner-most set is a list of old variable names and new variable names. The form of the RENAME= data set option is:

```
Data-Set-Name(rename=(Old-name1 = New_Name1 Old_Name2 = New_Name2 . .
.));
```

> You can rename as many variables as needed using the RENAME= data set option. However, the two variables on each side of the equal sign must be the same type (either character or numeric) and length.

If you have a situation where the two variables you would like to use in a merge are not the same type, you have to do a bit more work. The solution to this problem is discussed in the next section.

Merging Two Data Sets with One Character and One Numeric BY Variable

Because you cannot use a RENAME= data set option with two variables of different types, you need to create a new variable with the same name and type as the matching variable in the other data set. In this example, you have a variable called SS (Social Security number) in both files; however, one is stored as a character string and the other as a numeric value. Here is a listing of the two data sets you want to merge:

Listings of Data Set One_Char and Two_Num

Listing of Data Set ONE_CHAR

SS	Gender	Age
123-45-6789	M	45
088-54-1950	F	23
321-43-7766	M	68

Listing of Data Set TWO_NUM

Visit_Date	Fee_Paid	SS
10/14/2015	Yes	123456789
02/10/2015	No	88541950
03/23/2015	Yes	321437766

The program to create these two data sets is listed next, in case you want to try it out for yourself:

Program 14.21: Program to Create Data Sets One_Char and Two_Num

```
data One_Char;
   informat SS $11. Gender $1.;
   input SS Gender Age;
datalines;
123-45-6789 M 45
088-54-1950 F 23
321-43-7766 M 68
;
data Two_Num;
   informat Visit_Date mmddyy10. Fee_Paid $3.;
   input SS Visit_Date Fee_Paid;
   format Visit_Date mmddyy10.;
datalines;
123456789 10/14/2015 Yes
088541950 2/10/2015 No
321437766 3/23/2015 Yes
;
```

Before you conduct a merge, it is a good idea to run PROC CONTENTS (or examine the variables using SAS Studio) to ensure that the BY variables are the same type. With this in mind, here is a section of PROC CONTENTS output for each of the two data sets:

Output from PROC CONTENTS for Data Set One_Char

Alphabetic List of Variables and Attributes				
#	Variable	Type	Len	Informat
3	Age	Num	8	
2	Gender	Char	1	$1.
1	SS	Char	11	$11.

Output from PROC CONTENTS for Data Set Two_Num

Alphabetic List of Variables and Attributes					
#	Variable	Type	Len	Format	Informat
2	Fee_Paid	Char	3		$3.
3	SS	Num	8		
1	Visit_Date	Num	8	MMDDYY10.	MMDDYY10.

You can see that the variable SS is stored as character in data set One_Char and numeric in data set Two_Num.

It's time to merge these two data sets. You have a choice—convert the character variable to a numeric variable or vice versa. Let's choose the latter—changing the numeric variable to character. Here is the program:

Program 14.22: Merging Two Data Sets with One Character and One Numeric BY Variable

```
data Two_Char;
   set Two_Num(rename=(SS = SS_Num));
   SS = put(SS_Num,SSN11.);
   drop SS_Num;
run;

proc sort data=One_Char;
   by SS;
run;

proc sort data=Two_Char;
   by SS;
run;

Data One_and_Two;
   merge One_Char Two_Char;
   by SS;
run;

title "Listing of Data Set ONE_AND_TWO";
proc print data=One_and_Two noobs;
run;
```

There is quite a lot going on here. To start, you need to create a new data set (called Two_Char in this example) with a variable called SS that is stored as a character value. You use a SET statement to read the observations from data set Two_Num. Because you need the final character variable to be called SS, you use the RENAME= data set option to rename the original, numeric variable SS to SS_Num.

The next step is to use the PUT function to perform a numeric-to-character conversion. Here's how it works: There is a SAS format called SSN11. that writes out numeric representations of Social Security numbers as 11-byte character strings, complete with leading zeros and dashes. The PUT function takes two arguments. The first argument is the variable you want to convert (SS_Num in this example), and the second argument is a format that you want to use to format this value. The result is a character string in the same form as in data set One_Char. Because you no longer need (or want) the variable SS_Num in your data set, you use a DROP statement. This statement says that when SAS writes out observations to a data set, it should not include any variables listed on the DROP statement.

Output from Program 14.22 is listed next:

Output from Program 14.22

Listing of Data Set ONE_AND_TWO

SS	Gender	Age	Visit_Date	Fee_Paid
088-54-1950	F	23	02/10/2015	No
123-45-6789	M	45	10/14/2015	Yes
321-43-7766	M	68	03/23/2015	Yes

That wasn't easy, but it is very likely that some time in your programming career, you will face this problem. Now you know the solution.

Updating a Master File from a Transaction File (UPDATE Statement)

The final (yeah!) section of this chapter describes how to update values in a master file, using transaction data in a transaction file. In this example, you have a file of item numbers, descriptions, and prices, and you want to modify the file to change the price of a few items. Here is a program to create the master Price file and a listing of the file:

Program 14.23: Program to Create the Master File

```
data Price;
   input @1  Item_Number $5.
         @7  Description $10.
         @18 Price;
datalines;
12345 Hammer     11.98
22222 Saw        25.89
44010 Nails 10p  17.95
44008 Nails 8p   15.56
;

title "Listing of Data Set Price";
proc print data=Price;
   id Item_Number;
run;
```

Here is the listing:

Listing of Data Set Price

Item_Number	Description	Price
12345	Hammer	11.98
22222	Saw	25.89
44010	Nails 10p	17.95
44008	Nails 8p	15.56

You want to change the price of the hammer (item number 12345) to $12.98 and the price of 8 penny nails (item 44008) to $16.50. You first create a data set with these two items (you only need the item number and the new price). This will be your transaction data set.

Program 14.24: Creating the Transaction Data Set

```
data Transact;
   informat Item_Number $5.;
   input Item_Number Price;
datalines;
12345 12.98
44008 16.50
;

title "Listing of Data Set Transact";
proc print data=Transact;
   id Item_Number;
run;
```

This is the listing:

Output from Program 14.24

Listing of Data Set Transact

Item_Number	Price
12345	12.98
44008	16.50

Here is the program to update the prices in your master file:

Program 14.25: Updating Your Master File Using a Transaction Data Set

```
proc sort data=Price;
   by Item_Number;
run;
```

```
proc sort data=Transact;
   by Item_Number;
run;

data Price_10Oct2015;
   update Price Transact;
   by Item_Number;
run;

title "Listing of Data Set Price_10Oct2015";
proc print data= Price_10Oct2015;
   id Item_Number;
run;
```

You first sort both data sets by Item_Number. Next, you decide to give the new Price data set a different name—Price10Oct2015. Giving the new Price data set a new name is a good idea—it helps prevent confusion. Finally, you use an UPDATE statement to update the prices in the original Price data set. UPDATE is similar to MERGE. However, in a MERGE, if you have a variable in the second data set with the same name as a variable in the first data set, the value in the second data set will replace the value in the first data set, even if that value is a missing value. When you use the UPDATE statement, a missing value in the second data set does not replace the value in the first data set—just what you want to happen. Here is the listing:

Output from Program 14.25

Listing of Data Set Price_10Oct2015

Item_Number	Description	Price
12345	Hammer	12.98
22222	Saw	25.89
44008	Nails 8p	16.50
44010	Nails 10p	17.95

You see that the two prices are updated. Keep the UPDATE statement in mind whenever you need to replace values in one data set using values from another data set. It is one of the often-forgotten SAS statements.

Conclusion

This was clearly one of the more difficult chapters. However, many, if not most, programming problems require you to manipulate data from multiple data sets. If you know and love SQL, you can use that knowledge to combine your SAS data sets. The choice of SET, MERGE, and UPDATE versus SQL is not always easy. We often use what we know best. Because PROC SQL came much later in the evolution of SAS, many of the "older" SAS programmers choose to use DATA step processing.

Problems

1. Starting with the Sashelp data set Fish, create a data set called Small_Perch that contains only perch that weigh less than 50 (whatever the weigh units are). Do this using a WHERE statement.
2. Repeat Problem 1 using a WHERE= data set option.
3. You are reading raw data from a data set created by Program for Problem Sets 1 (listed below). Use a subsetting IF statement to include only those subjects where the sum of Q1–Q3 is greater than or equal to 6.

```
data Questionnaire;
   informat Gender 1. Q1-Q4 $1. Visit date9.;
   input Gender Q1-Q4 Visit Age;
   format Visit date9.;
datalines;
1 3 4 1 2 29May2015 16
1 5 5 4 3 01Sep2015 25
2 2 2 1 3 04Jul2014 45
2 3 3 3 4 07Feb2015 65
;
```

4. Starting with the Sashelp data set Cars, create two temporary data sets. The first one (Cheap) should include all the observations from Cars where the MSRP (manufacturer's suggested retail price) is less than or equal to $11,000. The other (Expensive) should include all the observations from Cars where the MSRP is greater than or equal to $100,000. Use a KEEP= data set option to include only the variables Male, Type, Origin, and MSRP from the Cars data set. Be sure to create these two data sets in one DATA step. Use PROC PRINT to list the observations in Cheap and Expensive. Even though there are no missing values for the variable MSRP, write your program so that any observation with a missing value for MSRP will not be written to data set Cheap.
5. Run Program for Problem Sets 6 to create two data sets (FirstQtr and SecondQtr). Then create a new data set (FirstHalf) that contains all the observations from FirstQtr and SecondQtr.

Program for Problem Sets 1

```
data FirstQtr;
   input Name $ Quantity Cost;
datalines;
Fred 100 3000
Jane 90 4000
April 120 5000
;
data SecondQtr;
```

```
   input Name $ Quantity Cost;
datalines;
Ron 200 9000
Jan 210 9500
Steve 177 5400
;
```

6. Repeat Problem 5, except use PROC APPEND to combine the observations from data sets FirstQtr and SecondQtr. Because you want the resulting data set to be called FirstHalf, you will first need to make a copy of FirstQtr that is called First_Half.
7. Run Program for Problem Sets 7. Then create a new data set (Both) that contains ID, X, Y, Z, and Name. Include only those IDs that are in both data sets.

Program for Problem Sets 2

```
data First;
   input ID $ X Y Z;
datalines;
001 1 2 3
004 3 4 5
002 5 7 8
006 8 9 6
;
data Second;
   input ID $ Nane $;
datalines;
002 Jim
003 Fred
001 Susan
004 Jane
;
```

8. Repeat Problem 7, except include all observations from each data set, even if there is no corresponding ID in one of the files.
9. Run Program for Problem Sets 8. Write a program to create a new data set called New_Prices, where the price of item X200 is $410 and the price of item A123 is $121. Caution: Item _Number is a character variable and the data set is not sorted by Item_Number.

Program for Problem Sets 3

```
data Prices;
   input Item_Number $ Price;
datalines;
A123 $123
B76 4.56
X200 400
D88 39.75
;
```

Chapter 15: Describing SAS Functions

Introduction

You have already seen a number of SAS functions in earlier chapters of this book. This chapter explores some of the most useful SAS functions that work with numeric and character data.

To review, SAS functions either return some system value or perform some calculation and return a value. Here are some useful facts about SAS functions:

- All SAS function names are followed by zero or more arguments, placed in parentheses following the function name.
- Although there are some exceptions, arguments to SAS functions can be variable names (the most common type of argument), constants (numbers for numeric functions, character values in quotes for character arguments), expressions (such as arithmetic expressions), or even other functions.
- Functions can only return a single value and, in most cases, the arguments to SAS functions do not change after you execute the function.

You will also see two CALL routines in this chapter. CALL routines are something like functions except for two important differences: One, some or all of the arguments can change value after the call; and two, you do not use a CALL routine in an assignment statement.

The following examples, using the WEEKDAY function (which returns the day of the week given a SAS date), demonstrate the variety of arguments that are valid with most SAS functions:

- `Day = weekday(Date)` where Date is the name of a SAS variable
- `Day = weekday(20013)` a numeric constant
- `Day = weekday('01Jan2015'd)` a date constant
- `Day = weekday(today())` where Today is a SAS function that returns the current date
- `Day = weekday(Date + 1)` an arithmetic expression giving the day of the week one day after Date

Describing Some Useful Numeric Functions

Each function in this chapter will be described and you will see either a full program or a statement showing how the function works.

Function Name: MISSING

What it does: The MISSING function returns a value of true (1) if the argument is a missing value and false (0) otherwise.

Arguments: A character or numeric value

Examples:

Program 15.1: Demonstrating the MISSING Function

```
data Old_Miss;
   input ID $ Age;
   if missing(Age) then Age_group = .;
   else if Age le 50 then Age_group = 1;
   else Age_group = 2;
datalines;
001 15
002 .
003 78
004 26
;

title "Listing of Data Set Old_Miss";
proc print data=Old_Miss noobs;
run;
```

Explanation: You use the MISSING function to test if Age is a missing value. If so, you assign a missing value to the variable Age_Group. Here is the listing from Program 15.1:

Output from Program 15.1

Listing of Data Set Old_Miss

ID	Age	Age_group
001	15	1
002	.	.
003	78	2
004	26	1

Notice that Age_Group has a missing value for ID 002.

Function Name: N

What it does: Returns the number of nonmissing values in the list of arguments

Arguments: One or more numeric values. If any of the arguments are in the form *Variable*1-*VariableN*, the list must be preceded by the keyword OF.

Examples:

```
X1=1;       X2=.;       X3=3;       Age=27;      Wt=.;       Ht=68;
```

```
① n(of X1-X3) = 2
② n(Age, Wt, Ht) = 2
```

Explanation: ① In the list of variables X1–X3, there are two nonmissing values.

② Among the three variables Age, Wt, and Ht, there are two nonmissing values.

Function Name: NMISS

What it does: Returns the number of missing values in the list of arguments

Arguments: One or more numeric values. If any of the arguments are in the form *Variable*1-*VariableN*, the list must be preceded by the keyword OF.

Examples:

```
X1=1;       X2=.;       X3=3;       Age=27;      Wt=.;       Ht=68;
```

```
① nmiss(of X1-X3) = 1
② nmiss(Age, Wt, Ht) = 1
```

Explanation: ① In the list of variables X1–X3, there is one missing value.

② Among the three variables Age, Ht, and Wt, there is one missing value.

Function Name: SUM

What it does: Returns the sum of the arguments. In performing this operation, missing values are ignored. If all the arguments are missing, the result is a missing value.

Arguments: One or more numeric values. If any of the arguments are in the form *Variable*1-*VariableN*, the list must be preceded by the keyword OF.

Examples:

```
X1=1;       X2=.;       X3=3;       Age=27;      Wt=.;       Ht=68;
```

```
① sum(of X1-X3) = 4
```

② **sum**(Age, Wt, Ht) = 95

Explanation: ① Because X2 is a missing value, the result is the sum of 1 and 3.

② The sum of Age (27) and Ht (68) is 95. The missing value is ignored.

Because the SUM function returns a missing value when all of its arguments are missing values, a popular trick to return a value of 0 in this situation is to include a 0 as one of the arguments. For example,

If Cost1–Cost5 are all missing, the expression

```
sum(0,of Cost1-Cost5)
```

will return a value of 0.

Function Name: MEAN

What it does: Returns the mean (average) of the arguments. In performing this operation, missing values are ignored. If all the arguments are missing, the result is a missing value.

Arguments: One or more numeric values. If any of the arguments are in the form *Variable*1-*VariableN*, the list must be preceded by the keyword OF.

Examples:

```
X1=1;       X2=.;       X3=3;       Age=27;       Wt=.;       Ht=68;
```

① **mean**(of X1-X3) = 2
② **mean**(Age, Wt, Ht) = 47.5

Explanation: ① X2 is a missing value and is ignored. The mean is computed as (1 + 3) / 2.

② The mean of Age (27) and Ht (68) is 47.5. The missing value is ignored.

Function Name: MIN

What it does: Returns the smallest nonmissing value of its arguments. If all the arguments are missing, the result is a missing value.

Arguments: One or more numeric values. If any of the arguments are in the form *Variable*1-*VariableN*, the list must be preceded by the keyword OF.

Examples:

```
X1=1;      X2=.;      X3=3;      Age=27;      Wt=.;      Ht=68;
```

```
① min(of X1-X3) = 1
② min(Age, Wt, Ht) = 27
```

Explanation: ① 1 is the smallest nonmissing value of the arguments.

② 27 is the smallest nonmissing value.

Function Name: MAX

What it does: Returns the largest nonmissing value of its arguments. If all the arguments are missing, the result is a missing value.

Arguments: One or more numeric values. If any of the arguments are in the form *Variable*1-*VariableN*, the list must be preceded by the keyword OF.

Examples:

```
X1=1;      X2=.;      X3=3;      Age=27;      Wt=.;      Ht=68;
```

```
① max(of X1-X3) = 3
② max(Age, Wt, Ht) = 68
```

Explanation: ① 3 is the largest value of the arguments.

② The largest value is 68.

Function Name: SMALLEST

What it does: Returns the *n*th smallest nonmissing value of its arguments. If there is no *n*th nonmissing value, the function returns a missing value.

Arguments: If the first argument is a 1, the function returns the smallest (nonmissing) value in the list of numeric values; if it is a 2, the function returns the second smallest value; and so forth. The second to last arguments are numeric values. If any of the arguments are in the form *Variable*1-*VariableN*, the list must be preceded by the keyword OF.

Examples:

```
X1= 1;      X2=.;      X3=3;      X4=2;
```

```
① smallest(1,of X1-X4) = 1
② smallest(2,of X1-X4) = 2
③ smallest(3,of X1-X4) = 3
④ smallest(4,of X1-X4) = . (missing value)
```

Explanation: ① When the first argument is equal to 1, the result is identical to the MIN function.

② When it is equal to 2, it returns the second (nonmissing) smallest number.

③ The third nonmissing smallest value is 3.

④ When you are asking for the fourth smallest number, the function returns a missing value because there is no fourth smallest number.

Function Name: LARGEST

What it does: Returns the *n*th largest nonmissing value of its arguments. If there is no *n*th nonmissing value, the function returns a missing value.

Arguments: If the first argument is a 1, the function returns the largest (nonmissing) value in the list of numeric values; if it is a 2, the function returns the second largest value; and so forth. The second through last arguments are numeric values. If any of the arguments are in the form *Variable*1-*VariableN*, the list must be preceded by the keyword OF.

Examples:

```
X1= 1;        X2=.;        X3=3;        X4=2;
```

```
① largest(1,of X1-X4) = 3
② largest(2,of X1-X4) = 2
③ largest(3,of X1-X4) = 1
④ largest(4,of X1-X4) = . (missing value)
```

Explanation: ① When the first argument is equal to 1, the result is identical to the MAX function.

② When it is equal to 2, it returns the second (nonmissing) largest number.

③ The third largest number is 1.

④ When you are asking for the fourth largest number, the function returns a missing value because there is no fourth largest number.

Programming Example Using the N, NMISS, MAX, LARGEST, and MEAN Functions

The program that follows demonstrates how several of the functions just described can work in concert to produce a very useful result. Here is the problem:

You are given the results of a psychological study. There are 10 variables, Q1 to Q10. You are asked to compute several scores as follows:

1. Score1: the mean of the first five questions (Q1–Q5). Compute this value only if there are three or more nonmissing values.
2. Score2: the mean of Q6–Q10. Compute this mean if there are two or fewer missing values.
3. Score3: the highest score of Q1–Q10.
4. Score4: the sum of the three highest scores of Q1–Q10.

Program 15.2: Program Demonstrating Several Functions (N, NMISS, MAX, LARGEST, and MEAN)

```
data Score;
   input ID $ Q1-Q10;
   if n(of Q1-Q5) ge 3 then Score1 = mean(of Q1-Q5);
   if nmiss(of Q6-Q10) le 2 then Score2 = mean(of Q6-Q10);
   Score3 = max(of Q1-Q10);
   Score4 = sum(largest(1,of Q1-Q10),
                largest(2,of Q1-Q10),
                largest(3,of Q1-Q10));
datalines;
001 9 7 8 6 7 6 . . 9 2
002 . . . . 9 8 7 8 9 9
003 6 7 6 7 6 . . . 9 9
;

title "Listing of Data Set Score";
proc print data=Score noobs;
run;
```

Explanation: The combination of N and MEAN or NMISS and MEAN is very useful for solving problems of this nature. Also, determining the second or third largest value in a list of values is very difficult without using the LARGEST function. Here is the output:

Output from Program 15.2

Listing of Data Set Score

ID	Q1	Q2	Q3	Q4	Q5	Q6	Q7	Q8	Q9	Q10	Score1	Score2	Score3	Score4
001	9	7	8	6	7	6	.	.	9	2	7.4	5.66667	9	26
002	.	.	.	.	9	8	7	8	9	9	.	8.20000	9	27
003	6	7	6	7	6	.	.	.	9	9	6.4	.	9	25

There are missing values for Score1 and Score2 because the criteria for computing a score were not met.

Function Name: INPUT

What it does: Takes the first argument (character value) and "reads" it as if it were being read from text data in a file using an INFORMAT that you supply as the second argument of the function. The most popular use of this function is to perform **character-to-numeric conversion**.

Arguments: The first argument is a character value, typically a character variable. The second argument is the informat you want to use to associate with the first argument.

Examples:

```
C_Num = '123';      C_Date = '10/21/1950';      C_Money = '$12,345.54';
```

```
Num = input(C_num,12.);  Num is the number 123
Date = input(C_Date,mmddyy10.);  Date is a SAS date
Money = input(C_Money,Dollar12.);  Money is a numeric variable
```

Explanation: In the first example, notice that the informat (12.) is much larger than you need (you only needed 3.). However, unlike reading data from a text file, the INPUT function will not read past the end of a character value, so there is no harm in choosing a large number for the numeric informat.

The second example shows how to convert a date, stored as a character string, into a true SAS date.

The last example uses the Dollar12. informat that strips dollar signs and commas from a value. Note that the Comma. informat also works for this example.

CALL Routine: CALL SORTN

What it does: After the CALL statement executes, the values of all of the calling arguments are in ascending order. That is, this CALL routine **can sort within an observation**.

Arguments: One or more numeric variables. If any of the arguments are in the form *Variable*1-*VariableN*, the list must be preceded by the keyword OF.

Examples:

```
X1=7;        X2=3;        X3=9;        X4=.;        X5=2;
```

```
call sortn(of X1-X5);
```

The values of X1 to X5 are now:

```
X1=.         X2=2         X3=3         X4=7         X5=9
```

```
call sortn(of X5 - X1);
```

The values of X1 to X5 are now:

```
X1=9         X2=7         X3=3         X4=2         x5=.
```

If you enter the arguments in reverse order, you can perform a descending sort of the values.

As a practical example, you have 10 test scores (Score1 to Score10) for each student (Stud_ID). You want to compute the mean of the eight highest scores.

Program 15.3: Using CALL SORTN to Compute the Mean of the Eight Highest Scores

```
data Test;
   input Stud_ID $ Score1-Score10;
   call sortn(of Score1-Score10);
   Mean_Top_8 = mean(of Score3-Score10);
datalines;
001 90 90 80 78 100 95 90 92 88 82
002 50 55 60 65 70 75 80 85 90 95
;

title "Listing of Data Set TEST";
proc print data=Test;
   id Stud_ID;
   var Score1-Score10 Mean_Top_8;
run;
```

Here is the listing from Program 15.3:

Output from Program 15.3

Listing of Data Set TEST

Stud_ID	Score1	Score2	Score3	Score4	Score5	Score6	Score7	Score8	Score9	Score10	Mean_Top_8
001	78	80	82	88	90	90	90	92	95	100	90.875
002	50	55	60	65	70	75	80	85	90	95	77.500

Explanation: Notice that the Score variables are now in ascending order. If you prefer, you can reverse the order of the arguments in the CALL SORTN routine and then compute the mean as `mean(of Score1-Score8).`

Function Name: LAG

What it does: If you execute the LAG function **for every iteration of the DATA step**, it returns the value of its argument from the previous observation.

> The true definition of the LAG function is that it returns the value of its argument the last time the function **executed**.

Arguments: A numeric variable

Examples:

Because SAS processes one observation at a time, you need special tools to be able to retrieve a value from an earlier observation. The LAG function is one of those tools. For this first example, you have daily stock prices, and you want to compare the current day's price with that of the previous day. Here is the program:

Program 15.4: Demonstrating the LAG Function

```
data Stocks;
   informat Date mmddyy10.;
   input Date Price;
   Up_Down = Price - lag(Price);
   format Date mmddyy10.;
datalines;
1/1/2015 100
1/2/2015 98
1/3/2015 96
1/4/2015 101
1/5/2015 101
1/6/2015 104
;

title "Listing of Data Set Stocks";
```

```
proc print data=Stocks noobs;
run;
```

Explanation: Because this program is executing the LAG function for every iteration of the DATA step, the variable Up_Down will be the current day's price minus the price from the previous day. Here is the listing:

Output from Program 15.4

Listing of Data Set Stocks

Date	Price	Up_Down
01/01/2015	100	.
01/02/2015	98	-2
01/03/2015	96	-2
01/04/2015	101	5
01/05/2015	101	0
01/06/2015	104	3

The value of Up_Down is missing in the first observation because that was the first time the LAG function executed and there was no previous value.

There is actually a whole family of LAG functions (LAG, LAG2, LAG3, etc.). LAG2 returns the value of its argument from two previous executions of the function, LAG3 three times, and so forth. As an example, you can use the Stocks data set to compute a three-day moving average as follows:

Program 15.5: Using the Family of LAG*n* Functions to Compute a Moving Average

```
data Moving;
   set Stocks;
   Yesterday = lag(Price);
   Two_Days_Ago = lag2(Price);
   Moving = mean(Price, Yesterday, Two_Days_Ago);
run;

title "Listing of Data Set MOVING";
proc print data=Moving noobs;
run;
```

Here is the output:

Output from Program 15.5

Listing of Data Set MOVING

Date	Price	Up_Down	Yesterday	Two_Days_Ago	Moving
01/01/2015	100	.	.	.	100.000
01/02/2015	98	-2	100	.	99.000
01/03/2015	96	-2	98	100	98.000
01/04/2015	101	5	96	98	98.333
01/05/2015	101	0	101	96	99.333
01/06/2015	104	3	101	101	102.000

Explanation: The variable Moving represents a three-day moving average. You may decide not to output this value until you reach day three—your choice.

Function Name: DIF

What it does: The value of DIF(X) is equal to X – LAG(X). Because one of the most common uses of the LAG function is to compute interobservation differences, SAS created the DIF function to save you a few keystrokes when you write your programs.

Arguments: A numeric variable

Examples:

You can substitute the line:

```
Up_Down = dif(Price);
```

for the line that computes the difference of the current day price and the price from the day before in Program 15.5.

Describing Some Useful Character Functions

The functions discussed in this section all deal with character values. Because there are so many useful character functions, you will see them grouped into logical categories, such as functions that extract substrings, functions that combine strings, and functions that take strings apart.

Function Names: LENGTHN and LENGTHC

What it does: LENGTHN returns the length of a character value, not counting trailing blanks (blanks to the right of the string). If the argument is a missing value, the function returns a 0. LENGTHC returns the storage length of a character variable.

Arguments: A character value or a character variable

Examples:

Program 15.6: Demonstrating the Two Functions LENGTHN and LENGTHC

```
data How_Long;
   length String $ 5 Miss $ 4;
   String = 'Abe';
   Miss = ' ';
   Length_String = lengthn(String);
   Store_String = lengthc(String);
   Display = ':' || String || ':';
   Length_Miss = lengthn(Miss);
   Store_Miss = lengthc(Miss);
run;

title "Listing of Data Set HOW_LONG";
proc print data=How_Long noobs;
run;
```

Explanation: This program demonstrates how the two functions LENGTHN and LENGTHC work. Looking at the output (below) should help clarify the difference between these two functions. The variable Display uses the concatenation operator (||) to place a colon on each side of String. This allows you to see any leading or trailing blanks in the value. Here is the output:

Output from Program 15.6

Listing of Data Set HOW_LONG

String	Miss	Length_String	Store_String	Display	Length_Miss	Store_Miss
Abe		3	5	:Abe :	0	4

The variable String is assigned a length of 5, using a LENGTH statement. The LENGTHN function returns a 3, the length of String with the (2) trailing blanks removed. The storage length, returned by the LENGTHC function, correctly shows that String has a length of 5. The variable Display is a colon, followed by 'Abe', followed by two blanks, followed by a colon.

Character missing values are represented by blanks. The number of blanks is equal to the storage length for that variable. However, when you are either assigning a missing value to a variable (as shown in the assignment statement for Miss) or testing if a character value is missing, you use a single blank in single or double quotes (you can also test for or assign a missing value by using two quotes together, with no space). The LENGTHN function returns a length of 0 for the variable Miss while LENGTHC returns the storage length (4).

For nonmissing character values, an older function called LENGTH is identical to the newer LENGTHN function. However, if the argument is a missing value, the older LENGTH function returns a 1 instead of a 0. The LENGTHN function was introduced with SAS version 9, the same version where strings of zero length were allowed. By the way, the 'N' in the function name LENGTHN stands for *null string*.

Function Names: TRIMN and STRIP

What it does: TRIMN removes trailing blanks and STRIP removes both leading and trailing blanks. Note that the TRIMN function replaces the older TRIM function. The difference is that the TRIMN function returns a null string (a string of zero length) if its argument is a missing value—the TRIM function returns a single blank when its argument is a missing value.

Both of these functions are particularly useful when you have character values that may contain leading or trailing blanks. Other functions that search strings for values such as digits or letters (see the ANY and NOT functions later in this chapter) search every position of a character value, including leading or trailing blanks. You usually want to remove leading and trailing blanks before using these searching functions.

Arguments: A character value

Examples:

```
Trail='ABC   ';      Lead= '   ABC';      Both='   ABC   ';
```

Expression	Value
`':' || Trail || ':'`	`:ABC   :`
`':' || trimn(Trail) ||`	`:ABC:`
`':' || Lead || ':'`	`:   ABC:`
`':' || strip(Lead) || ':'`	`:ABC:`
`':' || trimn(Both) || ':'`	`:   ABC:`
`':' || strip(Both) || ':'`	`:ABC:`

Explanation: The first example uses the concatenation operator to place a colon on each side of the variable Trail. Therefore, there are blanks between the 'C' and the colon.

In the second example, you use the TRIMN function to remove the trailing blanks before concatenating the final colon. Therefore, there are no blanks between the 'C' and the colon.

The third example concatenates a colon on each side of the variable Lead. Therefore, there are blanks between the first colon and the 'A'.

In the fourth example, the STRIP function removes leading and trailing blanks from the argument. Therefore, there are no blanks between the colons and 'ABC'.

In the fifth example, the TRIMN function removes the trailing blanks from Both so there are blanks between the first colon and the 'A' and no blanks between the 'C' and the final colon.

The last example uses the STRIP function to remove leading and trailing blanks from Both, resulting in no blanks between the first colon and the 'A' or the 'C' and the final colon.

> It is unusual for a SAS character value to contain leading blanks because the $*W*. informat left-justifies character values. However, if you are not sure if a variable contains leading blanks, use the STRIP function instead of the TRIMN function, just to be sure.

Before we leave these two functions, let's take a look at the following statements:

```
String = 'ABC   ';
String = trimn(String);
```

What is the value of String? The answer is '`ABC   `'. The reason is that although the TRIMN function removed the trailing blanks, when you then assign the trimmed value to a string of length 6, the trailing blanks return.

Function Names: UPCASE, LOWCASE, and PROPCASE (Functions That Change Case)

What it does: The three functions in this group all change the case of their argument. This is especially useful when you have character data that is entered in different cases (some upper, some lower, some mixed).

Arguments: LOWCASE and UPCASE: A character value. PROPCASE: A character value and, optionally, a second argument where you list your choice of delimiters.

You can probably guess the purpose of the first two functions, UPCASE and LOWCASE. UPCASE converts all letters in its argument to uppercase, and LOWCASE converts letters to lowercase. PROPCASE (stands for *proper case*) capitalizes the first letter of each "word" and converts the remaining letters to lowercase. The reason that "word" is in quotes is that besides the default blank delimiter, you can specify characters (in addition or instead of blanks) that you want to act as delimiters. The examples that follow will make this clear.

Examples:

```
Name1='rOn Cody';    Name2="D'amore"    String='AbC123xYZ';
```

```
upcase(Name1) = RON CODY
upcase(Name2) = D'AMORE
upcase(String) = ABC123XYZ

lowcase(Name1) = ron cody
lowcase(Name2) = d'amore
lowcase(String) = abc123xyz

propcase(Name1) = Ron Cody
propcase(Name2) = D'amore
propcase(Name2," '") = D'Amore
```

Explanation: The last two PROPCASE examples need some discussion. You probably want the letter following a single quote in a name to be displayed in uppercase. Because the default delimiter is a blank, using PROPCASE with a single argument results in the value D'amore. By including the second argument containing a blank and a single quote, the result is D'Amore. One final comment: Because you want to include a single quote in the list of delimiters, you need to use double quotes to enclose the second argument in this example.

There are some names that do not print properly after you use the PROPCASE function. For example, the name McDonald will become Mcdonald (lowercase 'd').

Function Name: PUT

What it does: Typically performs numeric-to-character conversion.

Arguments: First argument is a numeric or character value. The second argument is a format (either a built-in SAS format or one that you wrote). The PUT function takes the first argument, formats it using the second argument, and assigns the result to a character value.

Examples:

Program 15.7: Examples Using the PUT Function

```
proc format;
   value Agegrp low-50='Young'
                51-high='Older'
                     . ='Missing'
                 other ='Error';
run;

data Put_Eg;
   informat Date mmddyy10.;
   input SS_Num Date Age;
```

```
   SS = put(SS_Num,ssn11.);
   Day = put(Date,downame3.);
   Age_Group = put(Age,agegrp.);
   format Date date9.;
datalines;
123456789 10/21/1950 42
890001233 11/12/2015 86
987654321 1/1/2015 15
;
title "Listing of Data Set PUT_EG";
proc print data=Put_Eg noobs;
run;
```

Explanation: This example demonstrates several uses of the PUT function. You first want to create a character variable (SS) from the numeric variable SS_Num. The built-in SAS format SSN11. formats a numeric value in Social Security form (i.e., adds leading zeros and dashes). Next, by using the Downame3. format (Downame formats SAS dates to the names for the days of the week), you create a character variable containing the first three letters for each of the days. Finally, you create a format that groups Age into categories, and then you use the PUT function to create the variable Age_Group. This variable has values of 'Young' and 'Older' (plus 'Error' and 'Missing' if there are missing values and errors for Age). You would normally drop the variable SS_Num, keeping only SS, but it was not dropped so that you can see the original values of SS_Num in the listing.

Output from Program 15.7

Listing of Data Set PUT_EG

Date	SS_Num	Age	SS	Day	Age_Group
21OCT1950	123456789	42	123-45-6789	Sat	Young
12NOV2015	890001233	86	890-00-1233	Thu	Older
01JAN2015	987654321	15	987-65-4321	Thu	Young

Function Name: SUBSTRN (Newer Version of the SUBSTR Function)

What it does: Extracts a substring from a string.

Arguments: The first argument is a character value (one for which you want to extract a substring). The second argument is the starting position in the character value where you want to begin the substring. The third argument is the length of the substring you wish to extract. This last argument is optional—if you leave it out, the function returns characters from the starting position to the last non-blank character in the string.

Examples:

```
ID='123NJ456';    Amount='$12,456';
```

```
substrn(ID,4,2) = NJ
substrn(ID,4) = NJ456
substrn(Amount,2) = 12,456
substr(ID,4,2) = NJ
```

Explanation: In the first example, you want to extract the state codes from the variable ID. The starting position is 4 and you want to extract two characters. In the second example, you leave off the third argument (the length) and the SUBSTRN function returns all the remaining characters in the string.

The variable Amount starts with a dollar sign, and you want to extract the digits (and commas) without the dollar sign.

Finally, the older SUBSTR function, used in this example, returns the same substring as the SUBSTRN function. There are some more advanced features that are available to you with the SUBSTRN function compared to the SUBSTR function, but these features are not used in most programs.

> Very important point: If you do not define the length of the result of the SUBSTRN function, SAS will give it a default length equal to the length of the first argument (you cannot extract a substring longer than the original string). Every time you create a new variable using the SUBSTRN function, go back to the top of the DATA step and add a LENGTH statement, specifying the length of the substring variable.

Function Names: FIND and FINDC

What it does: FIND searches a string for a specific string of characters and returns the position in the string where this substring starts. If the substring you are searching for is not found, FIND returns a 0.

FINDC searches a string for any one of the characters you specify. It returns the position of the first character that it finds. As with the FIND function, if the search fails, the function returns a 0.

Arguments: The first argument in both of these functions is the character value you want to search for. For the FIND function, the second argument is the substring you are looking for. For the FINDC function, the second argument is a list of individual characters you are searching for. Both functions allow a third, optional argument (called a *modifier*) that allows these two functions to ignore case when doing their search.

There is also an optional third or fourth argument where you can specify a starting position for the search. If you use a modifier and a starting position, it doesn't matter what order you place these two arguments. If you only use a modifier or a starting position, place it as the third argument. SAS figures out which is a modifier and which is a starting position because modifiers are always character values and starting positions are always numeric values.

Examples:

```
String1='Good bad good';    String2='NNyYnYNN';
```

```
① find(String1,'good') = 10 (position of 'good' - lowercase)
② find(String1,'good','i') = 1 (the 'i' modifier says to ignore case)
③ find(String1,'ugly','i') = 0 (did not find string 'ugly')

④ findc(String2,'Y') = 4 (position of first uppercase 'Y')
⑤ findc(String2,'Y','i') = 3 (the 'i' modifier says to ignore case)
⑥ findc(String2,'X') = 0 (no 'X' in string)
⑦ findc(String1,'abcd') = 4 (position of 'd')
```

Explanation: ① FIND returns a 10 because the search is case-sensitive and the string 'good' starts in column 10.

② FIND uses the 'i' (ignore case) modifier, so it finds the string 'Good' in the first position.

③ There is no string 'ugly', so the function returns a 0.

④ FINDC is searching for an uppercase 'Y' and finds it in position 4.

⑤ The 'i' modifier is used and the lowercase 'y' in position 3 is chosen.

⑥ There is no 'x' in the string, so the function returns a 0.

⑦ You are looking for an 'a', 'b', 'c', or 'd'. The first character it finds in the string is the 'd' in position 4.

Note that these two functions replace the older INDEX and INDEXC functions (with added capability). There is also a third function, FINDW, that searches for words (strings bounded by word boundaries or other delimiters). Because this is not used as much as the other two functions, it was not included. You can find information on FINDW in the documentation.

Function Names: CAT, CATS, and CATX

What it does: You have already seen the concatenation operator (either || or !!) earlier in this chapter. Why do you need functions to concatenate strings? There are several advantages that these functions provide, as you will see in the examples that follow. CAT works in a similar manner to the concatenation operator. It takes a list of arguments and puts them together. CATS (pronounced Cat – S by most) is similar to CAT except that it strips leading and trailing blanks from each of the strings before putting them together (hence, the 'S' in the name). Finally, CATX does everything CATS does with one additional feature—it uses the first argument to the function as a separator between the strings. Note that the first argument to CATX can be more than one character.

> Very important point: If you assign the result of a concatenation function to a variable and you do not define a length for this variable, SAS will assign a length of 200. This is different from the rule for the concatenation operator. When you use the operator, the length of the result is the sum of the lengths of each of the strings you are concatenating.

Arguments: CAT and CATS: One or more character or numeric values. If you have variables in the form *Variable*1-*VariableN*, precede the list with the keyword OF. CATX: The first argument specifies one or more characters to use as separators between the strings. The second through last arguments are identical to the other CAT functions.

Examples:

```
Trail='ABC   ';      Lead= '   ABC';       Both='   ABC   ';
C1='A';    C2='B';   C3='C';    C4='D';    C5='E';
```

```
cat(':',Trail,Lead,':') = :ABC      ABC:
cats(':',Trail,Lead,':') = :ABCABC:
catx('-',':',Trail,Lead,':') = ABC-ABC:
catx('.',908,782,1323) = 908.782.1323
cats(of C1-C5) = ABCDE
```

Explanation: In the first example, you are concatenating a colon, the two variables Trail and Lead, and another colon. Notice that Trail has three trailing blanks and Lead has three leading blanks. The result has six blanks between the letters.

The second example uses the CATS function. Therefore, there are no spaces between the letters.

In the third example, the first argument is a dash, so the resulting string is similar to the CATS example except you now have a dash between the letters.

The next CATX example demonstrates that the arguments to all the CAT functions can be numeric. Remember that the resulting string is a character value.

The last example takes the five variables (C1–C5) and creates a single string.

In practice, the two functions CATS and CATX are used the most. Notice that in the fourth example, the arguments are numeric values. CAT, CATS, and CATX can all take either character or numeric arguments. When you use numeric arguments, an automatic conversion from numeric to character is performed and there are **no conversion messages in the log**.

You will see later how the CATS function and the COUNTC function make for a very powerful duo.

Function Names: COUNT and COUNTC

What it does: COUNT counts the number of substrings in a character value. COUNTC counts the number of individual characters in a character value.

Arguments: The first argument in both functions is a character value. For the COUNT function, the second argument is the substring you are searching for. For the COUNTC function, the second argument is a list of individual characters you want to count. As with the CAT functions, if you have a list of character variables in the form *Variable*1-*VariableN*, precede the list with the keyword OF. Both functions can take a third, optional argument, the most popular being 'i' that means ignore case (see the examples).

Examples:

```
String='Good bad good Good';    String2='AABBxxxcccc';
```

```
count(String,'good') = 1 (the search is case-sensitive)
count(String,'good','i') = 3 ('i' is the ignore case modifier)
countc(String2),'ABC') = 4 (2 A's and 2 B's - the c's are lowercase)
countc(String2,'ABC','i') = 8 (the 'i' modifier is specified)
```

Explanation: The first example is searching for the string 'good' and it only finds it once.

Because the 'i' modifier is used in the second example, the function finds three occurrences of 'good'.

In the first COUNTC example, there are a total of four uppercase A's and B's.

In the last COUNTC example, the 'i' modifier allows the four lowercase c's to be counted as well.

In the example shown next, you have five variables (Q1–Q5), and you want to count the number of uppercase or lowercase Y's in these five variables. The classical approach to solving this problem is to create an array of the five 'Q' variables, set a counter to 0, and use a DO loop to test each of the five 'Q' variables for an uppercase or lowercase 'Y'. Each time you find one, you increment the counter.

You can greatly simplify this problem by using the CATS function to place each of the five 'Q' values in a single character string. In the first observation, the result of the CATS function is the string 'YyYnn'. You use this as the argument of the COUNTC function to count the number of uppercase or lowercase Y's.

Program 15.8: Demonstrating the Combination of CATS and COUNTC

```
data Survey;
   input (Q1-Q5)($1.);
   Number_Y = countc(cats(of Q1-Q5),'Y','i');
datalines;
YyYnn
NNnnn
NYNyy
;
title "Listing of Data Set Survey";
proc print data=Survey noobs;
run;
```

Next time you find yourself creating an array and DO loops, consider if the problem can be solved using the method described in this example. The output from Program 15.8 is shown below:

Output from Program 15.8

Listing of Data Set Survey

Q1	Q2	Q3	Q4	Q5	Number_Y
Y	y	Y	n	n	3
N	N	n	n	n	0
N	Y	N	y	y	3

Imagine, a one-line solution to this problem!

Function Name: COMPRESS

What it does: Traditionally, it removes characters from a string. With the 'k' modifier, you can use this function to extract characters (for example, all digits) from a string. This is one of the most powerful character functions in the SAS arsenal—don't skip this section!

Arguments: If you provide only one argument (a character value), this function removes blanks from the string. An optional second argument is a string of characters that you want to remove from the first argument. If you provide an optional third argument, you can specify character classes to remove from the first argument (such as all letters or all digits).

One of the most useful applications of this function is to use a 'k' (keep) modifier as the third argument. When you do this, the characters you list as the second argument and/or the characters you specify with modifiers in the third argument are kept and all others are removed. The examples that follow should make this much clearer.

A list of some of the more useful modifiers is shown here:

Modifier	Description
'a'	All upper- and lowercase letters
'd'	All digits
'p'	All punctuation (such as periods, commas, etc.)
's'	All whitespace characters (spaces, tabs, linefeeds, carriage returns)
'i'	Ignore case
'k'	Keep the specified characters; remove all others (very useful)

Examples:

```
String1='abc def 123';   String2='(908)782-1234';   String3='120 Lbs.';
```

```
① compress(String1) = abcdef123 (one argument, default action remove
blanks)
② compress(String1,'0123456789') = abc def (remove digits 0-9)
③ compress(String1,,'d') = abc def ('d' modifier means remove all
digits)
④ compress(String2,,'kd') = 9087821234 (keep the digits)
⑤ compress(String3,,'kd') = 120 (this is a character string, not a
number)
⑥ input(compress(String3,,'kd')) = 120 (numeric value)
⑦ compress(String1,' ','kadp') = abc def 123
```

Explanation:

① Because there is only one argument, the COMPRESS function removes all blanks.

② The second argument (all the digits 0 to 9) are removed. Back in SAS 8, the COMPRESS function did not have a third (modifier) argument, and this was the way you would need to remove all digits. You may see examples like this if you are looking at programs written before SAS 9.

③ It is important to notice that there are two commas in a row following the first argument. With only one comma, the function would think that 'd' is the second argument and you are trying to remove all 'd's from the string. Using two commas tells the function that 'd' is the third argument (modifier) and you want to remove all digits.

④ One of my favorites—keep the digits and throw everything else away.

⑤ Again, keep the digits. Here it is used to remove units from a value.

⑥ This example shows how to combine the INPUT function and the COMPRESS function to extract the numeric value when there are units (or other non-digit values) included. The result of the COMPRESS function is a character value, and the INPUT function performs the character-to-numeric conversion.

⑦ If you have character data that includes non-printing ASCII or EBCDIC characters, you can use an expression similar to this. Here you are keeping spaces, uppercase and lowercase letters, digits and punctuation, and throwing everything else away.

To further illustrate the power of the COMPRESS function, the next program shows how to deal with a quantity that contains different units (pounds and kilograms) and wind up with numeric values, all in the same units:

Program 15.9: Using the COMPRESS Function to Read Data That Includes Units

```
data Weight;
   input Wt $ @@;
   Wt_Kg = input(compress(Wt,,'kd'),12.);
   if findc(Wt,'L','i') then Wt_Kg = Wt_Kg / 2.2;
datalines;
120lbs. 90Kg 80Kgs. 200Lb
;
title "Listing of Data Set WEIGHT";
proc print data=Weight noobs;
run;
```

You have weights that include units. Notice that the case of the units is not consistent and the units may or may not include periods. The double trailing @ on the INPUT statement allows you to read multiple observations on one line of data (it prevents the program from going to a new line each time that the DATA step iterates). You use the COMPRESS function with the 'kd' modifier to extract the digits from each weight and the INPUT function to perform the character-to-numeric conversion. Although you call this Wt_Kg, it may actually be in pounds. Next, you check to see if there is an upper- or lowercase 'L' in the units of the Wt variable with the FINDC function. If you find an 'L', you know the value is in pounds, so you divide Wt_Kg by 2.2 to convert the value to kilograms.

Here is the listing:

Output from Program 15.9

Listing of Data Set WEIGHT

Wt	Wt_Kg
120lbs.	54.5455
90Kg	90.0000
80Kgs.	80.0000
200Lb	90.9091

Function Name: SCAN

What it does: Takes a string apart (parses a string). The most common use is to extract the first and last name from a variable that contains the entire name (and possibly middle initial). The length of the result will be the same as the length of the first argument if you do not use a LENGTH statement to specify the length of the result.

Arguments: The first argument is the string you want to take apart. The second argument specifies which "word" you want. The reason "word" is in quotes is because if you do not list one or more delimiters in the third argument, the default action of the SCAN function is to use blanks plus other characters such as commas, periods, dashes, etc., as delimiters. In addition, the list of default delimiters is slightly different between ASCII and EBCDIC character sets. It is a good idea to specify exactly what delimiters you want as the optional, third argument. If the second argument is negative, the scan proceeds from right to left.

Examples:

```
String1='Alfred E. Newman';    String2='Cody, Ronald';    String3='12-34-56';
```

```
① scan(String1,1,' ') = Alfred (Specifying a blank delimiter)
② scan(String1,2,' ') = E. (The second word)
③ scan(String1,3) = Newman (Using the default delimiters that include
blanks)
④ scan(String1,-1,' ') = Newman (a negative second argument means
scan right to left)
⑤ scan(String1,4) = ' ' (missing value - there is no fourth word)
⑥ scan(String2,1,', ') = Cody (Specifying blanks and commas as
delimiters)
⑦ scan(String3,2,'-') = 34 (Specifying a dash as the delimiter)
```

Explanation:

① By specifying a blank delimiter, the first word is 'Alfred'.

② The second word is 'E'.

③ Because one of the default delimiters is a blank, the third word is 'Newman'. It's probably better to specify a blank as the delimiter if that is the only delimiter you want to use.

④ Using a -1 as the second argument is extremely useful in situations where you sometimes have first name and last name and at other times your name variable contains a middle name or initial. Using the -1 causes a right-to-left scan so you can easily extract the last name in either of these situations.

⑤ Here you are looking for the fourth word in a string that only contains three words, so the function returns a missing value. This provides you with a useful tool for extracting words from strings when you don't know how many words there are. You can extract each word in a DO loop and you know you are finished when the SCAN function returns a missing value.

⑥ Because blanks and commas are both used as delimiters in String2, you can specify both in the third argument. Any combination of commas and blanks will be treated as a single delimiter.

⑦ The SCAN function has many other uses besides extracting words from a string. By specifying the correct delimiter, you can parse other types of data.

When you use the SCAN function, you will probably want to use a LENGTH statement to declare a length for the result so that the result does not have the default length equal to the length of the first argument.

CALL Routine: CALL MISSING

What it does: Sets all the arguments (character and/or numeric) to a missing value after the CALL statement executes. This routine is especially useful when you are writing a program where you need to initialize a large number of character or numeric variables to a missing value.

Arguments: As many character and/or numeric variables as you want. If you have a variable list in the form *Variable*1-*VariableN*, precede the list by the keyword OF.

Examples:

```
X1=1;  x2=2;  x3=3;  C1='ABC';  C2='D';  C3='Fred';  C4='***';
```

```
call missing(of X1-X3, of C1-C4);
```

After the call, the variables X1–X3 are all numeric missing values and the variables C1–C4 are all character missing values.

Function Names: NOTDIGIT, NOTALPHA, and NOTALNUM

There are many more NOT functions (NOTSPACE, NOTPUNCT, etc.), but these three are the most useful.

What it does: Determines the first position in a string that is not a digit (NOTDIGIT), an upper- or lowercase letter (NOTALPHA), or a letter or digit (NOTALNUM). The functions all return the position of the first character in a string that is not in one of the specified categories. If all the characters in the string fit the category, the functions return a 0. One of the most useful applications of these functions is for data cleaning. If you have rules concerning allowable characters in a character variable (all digits, for example), you can use the appropriate NOT function to test if there are any characters that violate your condition.

Arguments: The first argument is a character value that you want to test. There is an optional second argument that is the starting position to begin the search. If the starting position is a negative number, go to the absolute value of the starting position and search from right to left.

Examples:

```
String1='1234NJ';    String2='abc123';    String3='123456';
```

```
notdigit(String1) = 5 (the position of the 'N')
notalpha(String2) = 4 (the position of the '1')
notalnum(String2) = 0 (all characters are alphanumeric)
notdigit(String3) = 0 (all characters are digits)
notalpha(String2,-3) = 0 (start searching at position 3 and search
right to left)
```

Explanation: In the first example, the 'N' is the first non-digit in the string. In the second example, the '1' is the first non-letter in the string. In the third example, you get a 0 because NOTALNUM returns the position of the first character that is not a letter or digit. Because String2 contains only digits and letters, the function returns a 0. String3 contains all digits. Therefore, NOTDIGIT returns a 0. In the last example, a starting position of -3 is an instruction to go to position 3 in String2 and search from right to left. There are only digits from the third position to the beginning of the string, so NOTALPHA returns a 0.

Function Names: ANYDIGIT, ANYALPHA, and ANYALNUM

What it does: Searches a string for the first occurrence of a digit (ANYDIGIT), an upper- or lowercase letter (ANYALPHA), or any digit or letter (ANYALNUM).

Arguments: The first argument is a character value you want to test. There is an optional second argument that is the starting position to begin the search. If the starting position is a negative number, go to the absolute value of the starting position and search from right to left.

Examples:

```
String1='1234NJ';    String2='abc123';    String3='123456';
```

```
anydigit(String2) = 4 (position of the '1')
anyalpha(String1) = 5 (position of the 'N')
anyalpha(String3) = 0 (no letters in the string)
anyalnum(String2) = 1 (position of the 'a')
```

Explanation: In the first example, the first digit in String2 is the '1' in position 4. In the second example, the 'N' is the first letter and it is in position 5. In the third example, String3 does not contain any letters and the function returns a 0. In the last example, the first character 'a' is a letter or digit, so the function returns a 1.

Function Name: TRANWRD

What it does: Performs a find-and-replace operation. One popular use of this function is to help with address standardization, converting words like "Street" to "St." and "Road" to "Rd."

Arguments: The three arguments to the TRANWRD function are: 1) the string you want to modify, 2) the 'find' string, and 3) the 'replace' string. Because the 'replace' string may be longer than the 'find' string, the default length for the result is 200. This fits with the general SAS rule that if the length returned by a function is longer than the length of the string argument, the default length will be 200.

Examples:

```
String1='123 First Street';    String2='Mr. Frank Jones Jr.';
```

```
tranwrd(String1,'Street','St.) = 123 First St.
tranwrd(String1,'Road','Rd.') = 123 First Street
tranwrd(String2,'Jr.',' ') = Mr. Frank Jones
tranwrd(String2,'Mr.',' ') =   Frank Jones (2 leading blanks)
```

Explanation: In the first example, "Street" is replaced by the abbreviation "St." In the second example, because there is no "Road" in String1, nothing is changed. In the third example, you are substituting a blank for "Jr. ", removing "Jr." from the string. In the last example, the resulting string has two leading blanks, one from converting "Mr." to a blank and the other is the original blank that was between "Mr." and "Frank."

Conclusion

Having a knowledge of the SAS functions described in this chapter can save you immense time and effort. Some of the more obscure or complex SAS functions, including Perl regular expressions, were not covered here. Please refer to SAS online help or one of the SAS Press books for information on these functions.

Problems

1. You have a SAS data set called Questionnaire2. It contains variables Subj (subject) and Q1–Q20 (numeric variables). Some of the values of Q1–Q20 may be missing values. Compute the following scores:

   Score1 is the mean of Q1–Q10. Only compute Score1 if there are seven or more nonmissing values of Q1–Q10.

   Score2 is the median of Q11–Q20. Compute Score2 only if there are five or fewer missing values in Q11–A20.

   Score3 is the largest value in variables Q1–Q10.

   Score4 is the sum of the two largest values in variables Q1–Q10.

   To test your program, run the program below to create the Questionnaire2 data set:

   Program for Problem Sets 1

   ```
   data Questionnaire2;
      input Subj $ Q1-Q20;
   datalines;
   001 1 2 3 4 5 1 2 3 4 5 1 2 3 4 5 1 2 3 4 5
   002 . . . . 3 2 3 1 2 3 4 3 2 3 4 3 5 4 4 4
   003 1 2 1 2 1 2 12 3 2 3 . . . . . . 4 5 5 4
   004 1 4 3 4 5 . 4 5 4 3 . . 1 1 1 1 1 1 1 1
   ;
   ```

2. Modify Program for Problem Sets 2 to compute two new variables as follows: Mean_3_Large is the mean of the three largest values from Q1 to Q10. Mean_3_Low is the mean of the three lowest nonmissing values in variables Q1–Q10. To ensure a good bulletproof solution to this problem, do not compute a mean if there are fewer than three nonmissing values for variables Q1–Q10. That is not the case with the current data set, but it could happen. (Remember that the MEAN function ignores missing values.)
3. Run Program for Problem Sets 10 to create the data set Char_Data. Create a new SAS data set called Num_Data that has the three variables Date, Weight, and Height, which are numeric. Format Date with the DATE9. format. You will need to rename the variables derived from Char_Date so that those names can be used in the new data set. Hint: The RENAME= option is:

   ```
   data-set-name(rename=(old1=new1 old2=new2, . . .));
   ```

   Program for Problem Sets 2

   ```
   data Char_Data;
   ```

```
   length Date $10 Weight Height $ 3;
   input Date Weight Height;
datalines;
10/21/1966 220 72
5/6/2000 110 63
;
```

4. Working with the data set Questionnaire2 (Problem 1), use CALL SORTN to compute the mean of the 12 highest scores in variables Q1–Q20. It is OK if some of the scores have missing values. You can perform a sort either in ascending order by listing the variables as Q1–Q20 or in descending order by listing the variables as Q20–Q1. Don't forget the keyword OF in the CALL routine.
5. Run Program for Problem Sets 11 to create data set Oscar (which will be used for several of the remaining problems in this section).

   Program for Problem Sets 3

```
data Oscar;
   length String $ 10 Name $ 20 Comment $ 25 Address $ 30
          Q1-Q5 $ 1;
   infile datalines dsd dlm=" ";
*Note: the DSD option is needed to strip the quotes from
 the variables that contain blanks;
   input String Name Comment Address Q1-Q5;
datalines;
AbC "jane E. MarPle" "Good Bad Bad Good" "25 River Road" y n N Y
Y
12345 "Ron Cody" "Good Bad Ugly" "123 First Street" N n n n N
98x "Linda Y. d'amore" "No Comment" "1600 Penn Avenue" Y Y y y y
. "First Middle Last" . "21B Baker St." . . . Y N
;
```

   Using the two length functions, compute the length of String, not counting trailing blanks and the storage length of String. Call these two variables L1 and L2.

6. Modify Program for Problem Sets 11 so that String is in uppercase and Name is in proper case. Use appropriate delimiters so that the name d'amore is spelled D'Amore.
7. Modify Program for Problem Sets 11 to create a new variable called Two_Three that contains the second and third characters in String. Be sure the length of Two_Three is 2.
8. Modify Program for Problem Sets 11 to include a new variable Yes_No. This variable should have a value of "Y" if the variable Comment contains the string "Good" (ignore case) and "N" if Comment does not contain the string "Good". Yes_No should be a missing value if Comment is a missing value. In addition, create another variable called Count_Y that represents the total number of Ys in variables Q1–Q5. Hint: Use the CATS function to concatenate Q1–Q5.
9. First, run Program for Problem Sets 12. Modify this program to create a new numeric variable called Height that is the height in inches, computed from the variable Ht. Note: 1 inch = 2.54 cm.

Program for Problem Sets 4

```
Data How_Tall;
   input Ht $ @@;
*Note: the @@ at the end of the INPUT statement allow you
 to place several observations on one line of data;
datalines;
65inches 200cm 70In. 220Cm. 72INCHES
;
```

10. Using Program for Problem Sets 11, create a new variable called Last_Name (length 10) that is the last name computed from the variable Name. Hint: Remember that a minus value representing which "word" you want causes the function to scan from right to left. Also, convert Name to proper case before computing Last_Name.
11. Using Program for Problem Sets 11, convert the variable Address so that the words "Street" and "Road" are replaced by their abbreviations "St." and "Rd."

Chapter 16: Working with Multiple Observations per Subject

Introduction

Because SAS processes one observation at a time, you need special programming tools when you want to analyze data where there are multiple observations per subject or any other grouping variable. For example, suppose you have a data set in which each observation represents a patient visit to a clinic. You record the patient ID, the date of the visit, and some health-related values. Some common programming tasks would include computing differences in values from visit to visit or differences between the first visit and the last visit for each patient. This chapter describes methods for solving problems of this type.

Useful Tools for Working with Longitudinal Data

Three useful programming tools for working with longitudinal data are first. and last. variables, the LAG and DIF functions, and retained variables. You will use all of these tools in the programs that follow.

Data set Clinic, shown below, is a collection of data on patients visiting a medical clinic. Some patients have only one visit—others have multiple visits. Data for each visit is stored in a separate observation. The following program creates the Clinic data set as well as a listing:

Program 16.1: Creating the Clinic Data Set

```
data Clinic;
   informat Date mmddyy10. PtNum $3.;
   input PtNum Date Height Weight Heart_Rate SBP DBP;
   format Date date9.;
datalines;
001 10/21/2015 68 190 68 120 80
001 11/25/2015 68 195 72 122 84
002 9/1/2015 72 220 76 140 94
003 5/6/2015 63 101 78 118 66
003 7/8/2015 63 106 76 122 70
003 9/1/2015 63 105 77 116 68
;
title "Listing of Data Set CLINIC";
proc print data=Clinic;
   id PtNum;
run;
```

Here is the listing:

Output from Program 16.1

Listing of Data Set CLINIC

PtNum	Date	Height	Weight	Heart_Rate	SBP	DBP
001	21OCT2015	68	190	68	120	80
001	25NOV2015	68	195	72	122	84
002	01SEP2015	72	220	76	140	94
003	06MAY2015	63	101	78	118	66
003	08JUL2015	63	106	76	122	70
003	01SEP2015	63	105	77	116	68

Patient 001 had two visits, patient 002 had one visit, and patient 003 had three visits.

Describing First. and Last. Variables

With this type of data, one of the first things you want to do is determine when you are processing the first observation and when you are processing the last observation for each patient. For this example, you want to first sort the data set by PtNum and Date. (Note: Sorting by date is not necessary here because visits are in date order for each patient. It is just a precaution in case a visit is out of order.) Next, you create a new data set (let's call it Clinic_New), as follows:

Program 16.2: Creating First. and Last. Variables

```
proc sort data=Clinic;
   by PtNum Date;
run;

data Clinic_New;
   set Clinic;
   by PtNum;
   file print;
   put  PtNum= Date= First.PtNum=  Last.PtNum=;
run;
```

The key here is to follow the SET statement with a BY statement. Doing this creates two temporary variables, First.PtNum and Last.PtNum. These variables are automatically dropped from the output data set—they only exist as the DATA step is processing. That is the reason that you use a PUT statement to examine these variables. Using PROC PRINT will not work because these two variables are temporary and are not in data set Clinic_New.

> The statement FILE PRINT is an instruction to send the results of the PUT statement to the RESULTS window, not to the SAS log, which is the default location.

Here is the output from Program 16.2:

Output from Program 16.2

```
PtNum=001 Date=21OCT2015 FIRST.PtNum=1 LAST.PtNum=0
PtNum=001 Date=25NOV2015 FIRST.PtNum=0 LAST.PtNum=1
PtNum=002 Date=01SEP2015 FIRST.PtNum=1 LAST.PtNum=1
PtNum=003 Date=06MAY2015 FIRST.PtNum=1 LAST.PtNum=0
PtNum=003 Date=08JUL2015 FIRST.PtNum=0 LAST.PtNum=0
PtNum=003 Date=01SEP2015 FIRST.PtNum=0 LAST.PtNum=1
```

The variable First.PtNum is equal to 1 for the first visit for each patient and 0 otherwise—the variable Last.PtNum is equal to 1 for the last visit for each patient and 0 otherwise. Let's see how you can use the First. and Last. variables in a program.

Computing Visit-to-Visit Differences

The first task is to compute visit-by-visit changes in the heart rate and the systolic and diastolic blood pressure for each patient. Here is the program:

Program 16.3: Computing Visit-to-Visit Differences in Selected Variables

```
proc sort data=Clinic;
   by PtNum Date;
run;

data Diff;
   set Clinic;
   by PtNum;
   if First.PtNum and Last.PtNum then delete;
   Diff_HR = dif(Heart_Rate);
   Diff_SBP = dif(SBP);
   Diff_DBP = dif(DBP);
   if not First.PtNum then output;
run;

title "Listing of Data Set DIFF";
proc print data=Diff;
   id PtNum;
run;
```

It makes no sense to compute differences for patients with only one visit. Patients with one visit have both variables, First.PtNum and Last.PtNum, equal to 1 (it is both the first and last visit), so you delete these observations. You use the DIF function to compute differences in heart rate, systolic blood pressure, and diastolic blood pressure for each patient (you can review the DIF function in Chapter 15). For the first visit for each patient (not including the first patient), you are actually computing the difference in each variable based on the **last value from the previous patient and the current value**. This is OK. You are only outputting observations for the second through last visit for each patient. You may be tempted to compute the difference values conditionally (that is, only for the second through last visits). This does not work. Remember that the DIF function returns the value of its argument the last time the function executed. If you do not execute the DIF function for the first visit for each patient, when you finally execute it, the function will return the value from the last time the function executed, giving you the last value from the previous patient.

Here is the output from Program 16.3:

Output from Program 16.3

Listing of Data Set DIFF

PtNum	Date	Height	Weight	Heart_Rate	SBP	DBP	Diff_HR	Diff_SBP	Diff_DBP
001	25NOV2015	68	195	72	122	84	4	2	4
003	08JUL2015	63	106	76	122	70	-2	4	4
003	01SEP2015	63	105	77	116	68	1	-6	-2

Each observation shows differences between the current visit and the previous visit.

Computing Differences Between the First and Last Visits

The next task is to compute the difference in the three variables (Heart_Rate, SBP, and DBP) from the first visit to the last visit. The primary tool to solve this problem is to use retained variables. As you (hopefully) remember, variables defined in assignment statements are automatically set to a missing value at the top of the DATA step. By using a RETAIN statement, you can instruct the program not to set these variables to a missing value—you are retaining them. This is one way to have the program "remember" a value from some previous observation. If you set a retained variable equal to a value from the first visit, then that retained value will be available when you are processing the last visit, allowing you to compute a difference score. The next program uses this strategy:

Program 16.4: Computing Differences Between the First Visit and the Last Visit

```
proc sort data=Clinic;
   by PtNum Date;
run;

data First_Last;
   set Clinic;
   retain First_Heart_Rate First_SBP First_DBP; ①
   by PtNum;
   if First.PtNum and Last.PtNum then delete; ②

   if First.PtNum then do; ③
      First_Heart_Rate = Heart_Rate;
      First_SBP = SBP;
      First_DBP = DBP;
   end;

   if Last.PtNum then do; ④
      Diff_HR = Heart_Rate - First_Heart_Rate;
      Diff_SBP = First_SBP - SBP;
      Diff_DBP = First_DBP - DBP;
```

```
      output;
   end;

run;

title "Listing of Data Set FIRST_LAST";
proc print data=First_Last;
   id PtNum;
run;
```

① You use a RETAIN statement for the three variables that will contain the value of heart rate, SBP, and DBP from the first visit.

② Delete patients with only one visit.

③ When you are processing the first visit for each patient, set each of the retained variables equal to the value of heart rate, SBP, and DBP, respectively.

④ When you reach the last visit for each patient, compute the difference of the current value minus the value from the first visit. Also, output an observation to data set First_Last.

Here is the listing:

Output from Program 16.4

Listing of Data Set FIRST_LAST

PtNum	Date	Height	Weight	Heart_Rate	SBP	DBP	First_Heart_Rate	First_SBP	First_DBP	Diff_HR	Diff_SBP	Diff_DBP
001	25NOV2015	68	195	72	122	84	68	120	80	4	-2	-4
003	01SEP2015	63	105	77	116	68	78	118	66	-1	2	-2

Each observation represents the last visit for each patient (who had more than one visit) and the differences between the first visit and the last visit.

Counting the Number of Visits for Each Patient

Another common task is to count the number of observations (visits, in this example) for each subject or other grouping variable. One of the most straightforward ways to accomplish this is to initialize a counter to 0 when you are processing the first visit for each patient, increment the counter for each visit, and output the patient number and visit count when you are processing the last visit for each patient. The next program uses this technique to create a data set of patient numbers and visit counts:

Program 16.5: Counting the Number of Visits for Each Patient

```
proc sort data=Clinic;
   by PtNum;
run;
```

```
data Counts;
   set Clinic;
   by PtNum;
   if First.PtNum then N_Visits=0;
   N_Visits + 1;
   if Last.PtNum then output;
run;

title "Listing of Data Set COUNTS";
proc print data=Counts;
   id PtNum;
run;
```

Because you follow the SET statement with a BY statement, you have the temporary variables First.PtNum and Last.PtNum at your disposal. When you are processing the first visit for each patient, you set the variable N_Visits equal to 0. Then, for every visit (observation), you use a SUM statement to add one to the counter.

Because it is so important, let's take a minute to review the properties of the SUM statement. If you had used an assignment statement like this:

```
N_Visits = N_Visits + 1;
```

The variable N_Visits would be set to a missing value for each iteration of the DATA step. A SUM statement differs from an assignment statement because it is in the form:

```
Variable + Expression;
```

There is no equal sign in this expression. There are three very important properties of a SUM statement:

- First, the *variable* is automatically retained.
- Second, the *variable* is initialized to 0.
- Third, if the *expression* results in a missing value, it is ignored.

The bottom line is that the SUM statement as used here is simply a counter.

When you are processing the last visit for each patient (Last.PtNum is true), you output an observation. This observation contains the heart rate, the systolic and diastolic blood pressure, and the number of visits for each patient. Here is the listing:

Output from Program 16.5

Listing of Data Set COUNTS

PtNum	Date	Height	Weight	Heart_Rate	SBP	DBP	N_Visits
001	25NOV2015	68	195	72	122	84	2
002	01SEP2015	72	220	76	140	94	1
003	01SEP2015	63	105	77	116	68	3

The variable N_Visits represents the number of visits for each patient.

Conclusion

You need special programming tools to analyze data for which you have multiple observations for each value of a BY variable. This chapter described some of these tools.

Problems

1. First run Program for Problem Sets 13 to create the Clinic data set. Next, write a DATA step that computes the visit-to-visit differences in Heart_Rate and Weight. Be sure to omit any patient with only one visit and only output an observation for the second through last visits for each patient. Careful, the data set is not sorted by Subj or Date.

   Program for Problem Sets 1

```
data Clinic;
   informat Date mmddyy10. Subj $3.;
   input Subj Date Heart_Rate Weight;
   format Date date9.;
datalines;
001 10/1/2015 68 150
003 6/25/2015 75 185
001 12/4/2015 66 148
001 11/5/2015 72 152
002 1/1/2014 75 120
003 4/25/2015 80 200
003 5/25/2015 78 190
003 8/20/2015 70 179
;
```

2. Using the Clinic data set from Problem 1, create a new data set (New_Clinic) that has one observation per subject. This one observation should include the number of visits for each subject (N_Visits) and the change in heart rate and weight from the first visit to the last visit. Omit any subject with only one visit.

3. Try to answer this question without running Program for Problem Sets 14 first. What is the value of Last_*x* in each of the five observations?

Program for Problem Sets 2

```
data Tricky;
   input x;
   if x gt 5 then Last_x = lag(x);
datalines;
6
7
2
10
11
;
```

4. What's wrong with this program (assume that you have just run Program for Problem Sets 13)?

```
1   data New;
2      set Clinic;
3      if first.Subj then First_Wt = Weight;
4      if last.Subj then Diff-Wt = Weight - First_Wt;
5   run;
```

Chapter 17: Describing Arrays

Introduction

Cody's law of SAS programming states that if you are writing a program and it is becoming very tedious, you should stop what you are doing and ask yourself, "Is there a non-tedious way to write this program?" The answer is often "yes." Perhaps there is a function that will save you time and effort, perhaps a SAS macro. SAS arrays are often the tool to save you huge amounts of time and effort writing your program. This chapter demonstrates the enormous power of SAS arrays.

What Is an Array?

A SAS *array* (with the exception of a temporary array) is a collection of SAS variables. Using the array name and a subscript, an array element can represent any one of the variables included in the array. The best way to understand how an array works is to look at a program that doesn't use arrays and then see how array processing can make the program much less tedious to write. So, with that in mind, let's look at a program that is just crying out for an array.

You have been given data on a group of subjects that includes age, height, weight, heart rate, systolic blood pressure, and diastolic blood pressure. Missing values were coded as 999 for each of these variables. Because SAS doesn't treat values of 999 as missing values, you need to convert every value of 999 to a SAS missing value. Here is a program that doesn't use arrays to solve this problem:

Program 17.1: Program to Convert 999 to a Missing Value (Without Using Arrays)

```
data Health_Survey;
   input ID $ Age Height Weight Heart_Rate SBP DBP;
   if Age = 999 then Age = .;
   if Height = 999 then Height = .;
   if Weight = 999 then Weight = .;
   if Heart_Rate = 999 then Heart_Rate = .;
   if SBP = 999 then SBP = .;
   if DBP = 999 then DBP = .;
datalines;
001 23 68 190 68 120 999
002 56 72 220 76 140 88
003 37 999 999 80 132 78
004 82 60 110 80 999 999
;
title "Listing of Data Set Health_Survey";
proc print data=Health_Survey noobs;
run;
```

Here is the output:

Output from Program 17.1

Listing of Data Set Health_Survey

ID	Age	Height	Weight	Heart_Rate	SBP	DBP
001	23	68	190	68	120	.
002	56	72	220	76	140	88
003	37	.	.	80	132	78
004	82	60	110	80	.	.

The program works fine, but it is tedious. Imagine if there were 30 or 40 variables that needed processing. The program below uses arrays to save you from writing so many lines of code:

Program 17.2: Rewriting Program 17.1 Using Arrays

```
data Health_Survey;
   input ID $ Age Height Weight Heart_Rate SBP DBP;
   array miss[6] Age Height Weight Heart_Rate SBP DBP;
   do i = 1 to 6;
```

```
      if miss[i] = 999 then miss[i] = .;
   end;
   drop i;
datalines;
001 23 68 190 68 120 999
002 56 72 220 76 140 88
003 37 999 999 80 132 78
004 82 60 110 80 999 999
;
title "Listing of Data Set HEALTH_SURVEY";
proc print data=Health_Survey noobs;
run;
```

You create an array using an ARRAY statement. Array names follow the same rules as SAS variable names. You follow the array name with a set of brackets. You are free to use either ( ), { }, or [ ] when defining an array. Because SAS functions all use regular parentheses, it is better to use either the curly brackets { } or the square brackets [ ] when defining an array. Most SAS documentation uses the curly variety—this author likes the square brackets. Take your pick.

You place the number of variables included in the array inside the brackets. In later examples, you will see that you can also use an asterisk instead of the number, and SAS will do the counting for you. Following the array name and the brackets, you list all the variables that you want to include in the array. These variables must be either all numeric or all character.

Once you have defined your array, you can use the array name followed by a number in the brackets (referred to as a *subscript*) anywhere in the DATA step. The array reference `miss[1]` is the same as writing the variable name Age, the array reference `miss[2]` is the same as writing the variable name Height, and so forth. This allows you to use a DO loop to process all of the variables in the array.

The first iteration of the DO loop is the following statement:

```
if miss[1] = 999 then miss[1] = .;
```

which becomes

```
if Age = 999 then Age = .;
```

Once the DO loop has finished, each of the variables in the array have been processed. You don't need (or want) to DO loop counter (i) in the output data set, so you use a DROP statement to remove it from the data set. The listing of data set Health_Survey is identical to the output from Program 17.1.

Describing a Character Array

You can create an array of character variables in much the same way that you did for numeric variables. If the variables in your array have already been defined as character variables elsewhere in the DATA step, you can use the same syntax that you used for numeric arrays. If you are creating the array at a point in the DATA step where the variables have not yet been defined, you can either precede the ARRAY statement with a LENGTH statement or, better yet, define the variables as character as well as defining a storage length right on the ARRAY statement. As an example, an ARRAY statement to define 10 character variables, all of length 2, is:

```
array chars[10] $ 2 Char1-Char10;
```

You follow the array name with a dollar sign and a length, followed by your list of variables.

As an example of a program that uses a character array, the following converts all of the character variables in the data set to uppercase:

Program 17.3: Converting Character Variables to Uppercase

```
data Uppity;
   informat Name $15. Q1-Q5 $1.;
   input Name Q1-Q5;
   array up[6] Name Q1-Q5;
   do i = 1 to 6;
      up[i] = upcase(up[i]);
   end;
   drop i;
datalines;
fred a B c D e
Sue a b c d D
;
title "Listing of Data Set UPPITY";
proc print data=Uppity noobs;
run;
```

Each of the variables, Name and Q1–Q5, are now in uppercase (see listing below):

Output from Program 17.3

Listing of Data Set UPPITY

Name	Q1	Q2	Q3	Q4	Q5
FRED	A	B	C	D	E
SUE	A	B	C	D	D

Performing an Operation on Every Numeric Variable in a Data Set

Suppose you want to repeat the process demonstrated in Program 17.1, replacing all values of 999 with a SAS missing value, except you want to do this for every numeric variable in a SAS data set. A very useful strategy is to use the keyword _NUMERIC_ for your variable list when you define your array. When used in a DATA step, _NUMERIC_ refers to all the numeric variables defined up to that point in the DATA step. You will see how this works in the program that follows. But first, you should know that you can use the keyword _NUMERIC_ in VAR, KEEP, DROP, and all statements that allow you to specify a list of variables, even in procedures.

Suppose you are given a SAS data set called Big that contains many numeric and character variables. Suppose, further, that the value 999 was used to represent missing values. The following program creates a new data set (Big_New) where all values of 999 are replaced by a SAS missing value:

Program 17.4: Converting 999 to a Missing Value for All Numeric Variables in a Data Set

```
data Big_New;
   set Big;
   array all_nums[*] _numeric_;
   do i = 1 to dim(all_nums);
      if all_nums[i] = 999 then all_nums[i] = .;
   end;
   drop i;
run;
```

There are some important features in this program. First, the ARRAY statement uses the keyword _NUMERIC_. Because the ARRAY statement follows the SET statement, the array all_nums is an array of all the numeric variables in data set Big. If you placed the ARRAY statement before the SET statement, the array would not contain any variables. To save you the time and effort of counting the number of numeric variables in data set Big, you place an asterisk in the brackets. But what value do you use in your DO loop? The DIM function (stands for *dimension*) takes an array name as its argument and returns the number of variables in the array. You can use the method in this example any time you need to perform an operation on every numeric variable in a SAS data set. If you need to perform an operation on every character variable in a data set, use the keyword _CHARACTER_ when you define your array.

Performing an Operation on Every Character Variable in a Data Set

This section describes a program to convert every character variable in a data set to uppercase:

Program 17.5: Converting Every Character Variable in a Data Set to Uppercase

```
data Big_New;
   Set Big;
   array all_chars[*] _character_;
   do i = 1 to dim(all_chars);
```

```
      all_chars[i] = upcase(all_chars[i]_);
   end;
   drop i;
run;
```

When this program executes, all the character variables will be in uppercase. This program and the previous program that operated on all numeric variables will save you immense time and effort whenever you need to operate on every numeric or character variable in a data set.

Converting a Data Set with One Observation per Subject into a Data Set with Multiple Observations per Subject

A fairly common programming problem is to transform or restructure a SAS data set. For example, suppose you are given the following SAS data set:

Program 17.6: Program to Create One Observation per Subject Data Set

```
data Wide;
   input Subj $ Wt1-Wt5;
datalines;
001 120 122 124 123 128
002 200 190 188 180 173
003 115 114 113 110 90
;
title "Listing of Data Set Wide";
proc print data=Wide noobs;
run;
```

Here is the listing:

Output from Program 17.6

Listing of Data Set WIDE

Subj	Wt1	Wt2	Wt3	Wt4	Wt5
001	120	122	124	123	128
002	200	190	188	180	173
003	115	114	113	110	90

You want to create a data set with five observations per subject like this:

Listing of data set THIN

Subj	Time	Weight
001	1	120
001	2	122
001	3	124
001	4	123
001	5	128
002	1	200
002	2	190
002	3	188
002	4	180
002	5	173
003	1	115
003	2	114
003	3	113
003	4	110
003	5	90

There are several ways to go about this. One is to use PROC TRANSPOSE. A straightforward DATA step approach using arrays will also do the trick. Although you can program this task without using arrays, the solution shown here is the standard way of solving this problem. First the program, then the explanation:

Program 17.7: Converting a Data Set with One Observation per Subject into a Data Set with Multiple Observations per Subject

```
data Thin;
   set Wide;
   array wt[5];
   do Time = 1 to 5;
      Weight = Wt[Time];
      output;
   end;
   drop Wt1-Wt5;
run;
title "Listing of data set THIN";
proc print data=Thin noobs;
run;
```

First, a word (or two) about the ARRAY statement: Notice that there are no variables listed after the array name. When you omit the list of variables, SAS uses the array name as the base and creates variables using this base and the numbers from 1 to the number of variables in the array. The ARRAY statement in this program is equivalent to:

```
array wt[5] Wt1-Wt5;
```

Because you want a variable called Time in the new data set, you use that as the variable for your DO loop counter. Inside the loop, you set a new variable that you call Weight equal to each of the Wt variables and output an observation. It is important that the OUTPUT statement is inside the DO loop. For each observation coming in from the Wide data set, you are creating five observations in the Thin data set. You no longer need the original Wt variables, so you drop them. The listing of this data was shown previously.

Converting a Data Set with Multiple Observations per Subject into a Data Set with One Observation per Subject

This section describes a restructuring in the reverse direction—going from a data set with many observations per subject to a data set with one observation per subject. You are probably more likely to go from one to many, as described in the previous section, but the reverse process is sometimes needed. This program is a bit trickier. If you need to do this, be sure to use the program listed here as the basis for your own program. Of course, there is always PROC TRANSPOSE to consider.

You can use the data set Thin, created in the previous section, as the starting point to create one that is identical to the original Wide data set. Once more, first the program, then the explanation:

Program 17.8: Converting a Data Set with Multiple Observations per Subject into a Data Set with One Observation per Subject

```
proc sort data=Thin;
   by Subj;
run;

data Wide;
   set Thin;
   by Subj;
   array Wt[5];
   retain Wt1-Wt5;
   if first.Subj then call missing(of Wt1-Wt5);
   Wt[Time] = Weight;
   if last.Subj then output;
   keep Subj Wt1-Wt5;
run;
```

You first sort data set Thin because you plan to follow the SET statement in the DATA step with a BY statement. In the previous chapter, you saw that this process creates two new variables: First.Subj and Last.Subj. You will need both of these variables to make this program work. When you are reading the first weight for each subject, you want to set the variables Wt1–Wt5 to a missing value. You need to do this because of the RETAIN statement. Why do you need to retain these five variables? Without the RETAIN statement, each of the variables Wt1–Wt5 would be set to missing at the top of the DATA step. The RETAIN statement is an instruction not to set these variables to missing. Initializing the variables Wt1–Wt5 is not necessary in this example. However, suppose data set Thin looked like this:

Data Set Thin with a Missing Observation for Subject 002

Listing of Data Set THIN

Subj	Time	Weight
001	1	120
001	2	122
001	3	124
001	4	123
001	5	128
002	1	200
002	2	190
002	3	188
002	5	173
003	1	115
003	2	114
003	3	113
003	4	110
003	5	90

There is a missing weight for subject 002 at time 4. If you had this situation and did not initialize Wt1–Wt5 to missing for each subject, the missing observation would wind up with the weight at time 4 for the previous subject. Initializing Wt1–Wt5 is just a good precaution.

Back to the program. You assign the value of Weight for each of the array variables. When all the variables Wt1–Wt5 have been given a value, you output a single observation. The resulting data set is identical to the original data set Wide shown previously.

Conclusion

Arrays can save you time and effort in writing programs. Remember to use the keywords _NUMERIC_ and _CHARACTER_ to define arrays when you want to perform an operation on all numeric variables or all character variables in a data set. Finally, you can use arrays to transform or restructure data sets.

Problems

1. Rewrite Program for Problem Sets 15 using arrays.

   Program for Problem Sets 1

```
data Prob1;
   length Char1-Char5 $ 8;
   input x1-x5 Char1-Char5;
   x1 = round(x1);
   x2 = round(x2);
   x3 = round(x3);
   x4 = round(x4);
   x5 = round(x5);
   Char1 = upcase(Char1);
   Char2 = upcase(Char2);
   Char3 = upcase(Char3);
   Char4 = upcase(Char4);
   Char5 = upcase(Char5);
datalines;
1.2 3 4.4 4.9 5 a b c d e
1.01 1.5 1.6 1.7 1.8 frank john mary jane susan
;
title "Listing of Data Set Prob1";
proc print data=Prob1 noobs;
run;
```

2. Starting with the Sashelp data set Fish, create a new, temporary data set (Inches) where each of the variables Height, Width, and Length1–Length3 are converted by dividing their values by 2.54. Use an array to do this.
3. Modify Program for Problem Sets 16 so that all values of 999 are converted to a SAS numeric missing value and all character values of 'NA' (uppercase or lowercase) are converted to a SAS character missing value. Even though there are only four numeric and three character variables, define the arrays with the keywords _NUMERIC_ and _CHARACTER_. Remember that the DIM function returns the number of variables in an array.

   Program for Problem Sets 2

```
data Missing;
   input w x y z C1 $ C2 $ C3 $;
datalines;
999 1 999 3 Fred NA Jane
8 999 10 20 Michelle Mike John
11 9 8 7 NA na Peter
;
```

4. Use the Sashelp data set Heart and create a new, temporary SAS data set UpHeart that contains all the variables from Sashelp.Heart, but convert all of the character variables to uppercase. Use the keyword _CHARACTER_ to define your array and the DIM function to define the upper limit of your DO loop. Because the Sashelp.Heart data set is so large, use the

data set option OBS=10 to read only the first 10 observations from this data set. Your SET statement should look like this:

```
set SASHELP.Heart(obs=10);
```

Chapter 18: Displaying Your Data

Introduction

The chapter describes how to use PROC PRINT to list all or part of a SAS data set. PROC PRINT can create decent-looking reports that include column labels and summary data. However, if you need a fancier report or want more control over its appearance, you can use PROC REPORT to produce your report. There is a tradeoff: PROC REPORT can produce more sophisticated, custom reports, but it is more complicated to use. As a matter of fact, there are entire books dedicated to PROC REPORT. For many applications, especially with the added tricks described in this chapter, you may find that the reports created by PROC PRINT will do just fine.

Producing a Simple Report Using PROC PRINT

This first example uses PROC PRINT to list the contents of one of the Sashelp data sets called Shoes. This first program does not use any options or statements—it is "bare bones."

Program 18.1: Demonstrating PROC PRINT without any Options or Statements

```
title  "Listing of Data Set SHOES";
title2 "In the SASHELP Library";
title3 "-----------------------------------------------------";

proc print data=sashelp.shoes;
run;
```

Three TITLE statements were added. You can include up to 10 TITLE statements in a program. Remember, all the TITLE statements remain in effect until you change them. If you write a new TITLE statement, say TITLE2 following the PROC PRINT statement in Program 18.1, the original second title line will be replaced by the new TITLE2 text, and all title lines higher than two will be deleted. That is, anytime you submit a new TITLE*n* statement, all title lines greater than *n* are deleted.

By default, PROC PRINT will list all variables and all observations. Also, the order of the variables will be the order that they are stored in the SAS data set. Here is the listing:

Output from Program 18.1 (edited to save space)

Listing of Data Set SHOES
In the SASHELP Library

Obs	Region	Product	Subsidiary	Stores	Sales	Inventory	Returns
1	Africa	Boot	Addis Ababa	12	$29,761	$191,821	$769
2	Africa	Men's Casual	Addis Ababa	4	$67,242	$118,036	$2,284
3	Africa	Men's Dress	Addis Ababa	7	$76,793	$136,273	$2,433
4	Africa	Sandal	Addis Ababa	10	$62,819	$204,284	$1,861
5	Africa	Slipper	Addis Ababa	14	$68,641	$279,795	$1,771
391	Western Europe	Sandal	Rome	3	$1,249	$4,611	$48
392	Western Europe	Slipper	Rome	13	$42,442	$132,283	$1,829
393	Western Europe	Sport Shoe	Rome	14	$9,969	$74,848	$549
394	Western Europe	Women's Casual	Rome	2	$19,964	$62,256	$954
395	Western Europe	Women's Dress	Rome	16	$106,676	$389,861	$3,160

The column labeled Obs is created by the procedure and it is an observation number. If you modify your data set (by sorting, adding, or deleting observations, for example), the observation number associated with a particular line of data may change. Most programmers want to replace the Obs column with a variable that identifies the observation, such as an ID or name.

Therefore, the next step is to replace the Obs column with a variable of your choice. You do this by including an ID statement. You can also select which variables to include in the listing by adding a VAR statement. Using a VAR statement also specifies the order of the columns in the report. Here is an example:

Program 18.2: Adding ID and VAR Statements to the Procedure

```
title  "Listing of Data Set SHOES";
title2 "In the SASHELP Library";
title3 "----------------------------------------";
```

```
proc print data=sashelp.shoes;
   id Region;
   var Product Stores Sales Inventory;
run;
```

As you can see, an ID statement and a VAR statement were added. Before we show the output, you should know about a shortcut that saves you from typing the variable names. If you right-click on the Shoes data set in SAS Studio, you will see a list of the variables in the data set. Hold down the Ctrl key and click on the variables you want in the VAR statement and they will be highlighted. You can then left-click on any of the highlighted variables and drag them over to your program. The order in which you Ctrl-click the columns determines their order in the VAR statement. It looks something like this:

Figure 18.1: Copying Variable Names from the Library Listing to Your Program

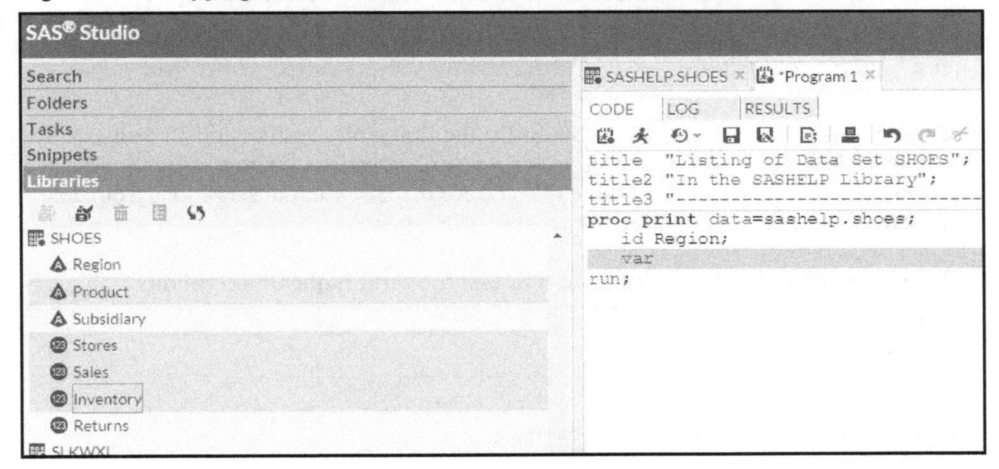

The variables Product, Stores, Sales, and Inventory were highlighted and dragged into the CODE window right next to the keyword VAR. The result is Program 18.2 listed above.

Here is the top portion of the revised listing:

Output from Program 18.2 (partial listing)

Listing of Data Set SHOES
In the SASHELP Library

Region	Product	Stores	Sales	Inventory
Africa	Boot	12	$29,761	$191,821
Africa	Men's Casual	4	$67,242	$118,036
Africa	Men's Dress	7	$76,793	$136,273
Africa	Sandal	10	$62,819	$204,284
Africa	Slipper	14	$68,641	$279,795

Region has replaced the Obs columns and only the variables listed in the VAR statement are included in the listing.

You can use a very useful data set option, OBS=, to list the first *n* observations in the data set. This is a very useful technique when you only want to see a few observations (from a possibly very big data set) to check that your program is working OK. To see the first eight observations from data set Shoes, you can add the OBS= option to the program, like this:

Program 18.3: Using the OBS= Data Set Option to List the First Eight Observations

```
title  "Listing of Data Set SHOES";
title2 "In the SASHELP Library";
title3 "------------------------------------------------";

proc print data=sashelp.shoes(obs=8);
   id Region;
   var Product Stores Sales Inventory;
run;
```

The data set option is placed in parentheses following the data set name. The listing will now show the first eight observations (shown below):

Output from Program 18.3

Listing of Data Set SHOES
In the SASHELP Library - First 8 Observations

Region	Product	Stores	Sales	Inventory
Africa	Boot	12	$29,761	$191,821
Africa	Men's Casual	4	$67,242	$118,036
Africa	Men's Dress	7	$76,793	$136,273
Africa	Sandal	10	$62,819	$204,284
Africa	Slipper	14	$68,641	$279,795
Africa	Sport Shoe	4	$1,690	$16,634
Africa	Women's Casual	2	$51,541	$98,641
Africa	Women's Dress	12	$108,942	$311,017

This listing stops at observation 8. If you want even more control over which observations to list, you can include the two data set options OBS= and FIRSTOBS=. You use FIRSTOBS= to specify the first observation to list and OBS= to specify the last observation to list. For example, to list observations 20 through 25, you would use:

```
proc print data=sashelp.shoes(firstobs=20 Obs=25);
```

You can use these data set options anywhere you specify a SAS data set name in a program or procedure.

Using Labels Instead of Variable Names as Column Headings

By default, PROC PRINT uses variable names, not variable labels, as column headings. If you would like variable labels to head your columns, use the LABEL procedure option. The program that follows creates a small data set (Dimensions) where several variables are given labels. You can then use PROC PRINT with and without the LABEL option to see its effect.

Program 18.4: PROC PRINT with and without a LABEL Option

```
data Dimensions;
   input Subj $ Ht Wt Waist;
   label Subj = 'Subject'
         Ht   = 'Height in Inches'
         Wt   = "Subject's Weight";
datalines;
001 68 180 35
```

```
002 75 220 40
003 60 101 28
;
title "PROC PRINT Without a LABEL Option";
proc print data=Dimensions;
   id Subj;
   Var Ht Wt Waist;
run;

title "PROC PRINT with LABEL Option";
proc print data=Dimensions label;
   id Subj;
   Var Ht Wt Waist;
run;
```

Before we look at the listings, take a look at the label for the variable Wt. Because you want to include a single quote in the label, you need to enclose the label in double quotes. In most places where you need to enclose a string in quotes, either a set of single or double quotes will work fine. This label is an exception. For more advanced programmers, you should also know that if you want SAS to resolve a macro variable, you need double quotes.

Here is the output:

Output from Program 18.4

PROC PRINT Without a LABEL Option

Subj	Ht	Wt	Waist
001	68	180	35
002	75	220	40
003	60	101	28

PROC PRINT with LABEL Option

Subject	Height in Inches	Subject's Weight	Waist
001	68	180	35
002	75	220	40
003	60	101	28

Each listing has a use. The first one (without the LABEL option) is most useful for a programmer, and the second one (with the LABEL option) is most useful for presenting the data to a non-programmer. Note that the variable Waist was not given a label, and its name is used in the second listing.

Including a BY Variable in a Listing

You can include a BY statement with PROC PRINT to break the output down by that variable. As an example, the program to produce the Health data set is reproduced here. To generate a listing broken down by Gender, you must first sort the data set by Gender and then include a BY statement with PROC PRINT. Here is the program:

Program 18.5: Listing the Health Data Set Broken Down by Gender

```
data health;
   infile '/folders/myfolders/health.txt' pad;
   input Subj   $ 1-3
         Gender $ 4
         Age      5-6
         HR       7-8
         SBP      9-11
         DBP      12-14
         Chol     15-17;
run;

proc sort data=Health;
   by Gender;
run;

title "Listing of Data Set HEALTH - by Gender";
proc print data=health;
   ID Subj;
   by Gender;
run;
```

The output looks like this:

Output from Program 18.5

Listing of Data Set HEALTH - by Gender

Gender=F

Subj	Age	HR	SBP	DBP	Chol
002	55	72	180	90	170
003	18	58	118	72	122
005	34	62	128	80	.
006	38	78	108	68	220

Gender=M

Subj	Age	HR	SBP	DBP	Chol
001	23	68	120	90	128
004	80	82	.	.	220

You now have separate listings for females and males.

Including the Number of Observations in a Listing

You can include a count of the number of observations in the entire data set or for a BY group by adding the option N='*label*' to PROC PRINT. To add a count to a listing of the Health data set, you could use the following program:

Program 18.6: Adding a Count of the Number of Observations to the Listing

```
title "Listing of Data Set HEALTH - by Gender";
proc print data=health n = "Total Observations =";
   ID Subj;
run;
```

Note: By adding the N= option to PROC PRINT, you now see the total number of observations in the listing, as shown next:

Output from Program 18.6

Listing of Data Set HEALTH

Subj	Gender	Age	HR	SBP	DBP	Chol
001	M	23	68	120	90	128
002	F	55	72	180	90	170
003	F	18	58	118	72	122
004	M	80	82	.	.	220
005	F	34	62	128	80	.
006	F	38	78	108	68	220
Total Observations = 6						

If you use the N= option along with a BY statement, you can see the observation count for each BY group (not shown).

Conclusion

There are several PROC PRINT options that are not described in this chapter. You can refer to the SAS Studio **Help** menu to see a complete list. However, if you need even more control over the format of your report, you may need to investigate PROC REPORT. I highly recommend the following book:

Carpenter, Art. 2007. *Carpenter's Complete Guide to the PROC REPORT Procedure*. Cary, NC: SAS Institute Inc.

Problems

1. Use PROC PRINT to list the first 10 observations in the Sashelp data set Fish. Replace the Obs column with the variable Species, and do not include the three variables Length1–Length3. Add three TITLE statements to produce the report heading as follows:

```
Listing of the First 10 Observations in Data Set Fish
Prepared by: (your name here)
-----------------------------------------------------
```

2. Prepare a report similar to the one in Problem 1 except break down the report by Species and include all the observations in the data set. Omit the three variables Length1–Length3 and use the PROC PRINT option NOOBS to omit the Obs column.
3. Repeat Problem 2 except use Species as a BY variable and as an ID variable. How does the listing differ from the listing in Problem 2?
4. Create a report from the Sashelp data set Retail. The report should include a title (of your choice) and show sales broken down by year. Include the number of observations for each year, and use variable labels instead of variable names as column headings. The beginning of your report should look like this:

YEAR	Retail sales in millions of $
1980	$220
	$257
	$258
	$295
Number=4	

YEAR	Retail sales in millions of $
1981	$247
	$292
	$286
	$323
Number=4	

Chapter 19: Summarizing Data with SAS Procedures

Introduction

This chapter describes how to summarize numeric data (means, standard deviations, etc.), and how to create SAS data sets containing this summary information. The main tools are PROC MEANS, PROC SUMMARY, and PROC UNIVARIATE.

Using PROC MEANS (with the Default Options)

PROC MEANS is one of the most useful procedures for summarizing data. As you will see in the programs that follow, you can run this procedure without specifying any options and obtain useful information, including the number of nonmissing observations, the mean, the standard deviation, and the minimum and the maximum values for all the numeric variables in the data set. Later on in this chapter, you will see how to add procedure options and statements to customize the summary report.

To demonstrate the various ways that you can use PROC MEANS to summarize data, run Program 19.1 to create a data set called Blood_Pressure. The program uses a random number generator to create a data set containing data from a fictitious drug trial.

The details of how this program works are not described in detail here, but some general information about this program may be interesting. The RAND function is used to generate random numbers that are normally distributed. The way the program is designed, if you run it yourself, you will obtain exactly the same data set used in the examples that follow. The program also generates a subject number, drug group, and gender, as well as the systolic and diastolic blood pressure (adjusted so that the drug group pressures are lower than the placebo group pressures). The program also outputs missing values for some of the variables, to make the data set more realistic. Here is the program:

Program 19.1: Creating the Blood_Pressure Data Set

```
libname Trial "/folders/myfolders";
data Trial.Blood_Pressure;
   call streaminit(37373);
   do Drug = 'Placebo','Drug A','Drug B';
      do i = 1 to 20;
         Subj + 1;
         if mod(Subj,2) then Gender = 'M';
         else Gender = 'F';
         SBP = rand('normal',130,10) +
               7*(Drug eq 'Placebo') - 6*(Drug eq 'Drug B');
         SBP = round(SBP,2);
         DBP = rand('normal',80,5) +
               3*(Drug eq 'Placebo') - 2*(Drug eq 'Drug B');
         DBP = round(DBP,2);
         Age = int(rand('normal',50,10) + .1*SBP);
         if Subj in (5,15,25,55) then call missing(SBP, DBP);
         if Subj in (4,18) then call missing(Gender);
         output;
      end;
   end;
   drop i;
run;

title "Listing of Data Set DRUG_TRIAL (first 10 observations)";
proc print data=Trial.Blood_Pressure(obs=10);
   id Subj;
run;
```

The first 10 observations of this data set are listed next:

Output from Program 19.1

Listing of Data Set DRUG_TRIAL (first 10 observtions)

Subj	Drug	Gender	SBP	DBP	Age
1	Placebo	M	138	86	50
2	Placebo	F	134	90	40
3	Placebo	M	136	84	61
4	Placebo		132	80	65
5	Placebo	M	.	.	71
6	Placebo	F	146	82	60
7	Placebo	M	138	88	70
8	Placebo	F	140	80	74
9	Placebo	M	134	80	59
10	Placebo	F	152	90	62

You are now ready to run PROC MEANS (without specifying any options) on this data set.

Program 19.2: Running PROC MEANS on the Blood_Pressure Data Set (Using All the Default Options)

```
title "Running PROC MEANS with all the Defaults";
proc means data=Trial.Blood_Pressure;
run;
```

After running this program, the RESULTS window looks like this:

Output from Program 19.2

Running PROC MEANS with all the Defaults

The MEANS Procedure

Variable	N	Mean	Std Dev	Minimum	Maximum
Subj	60	30.5000000	17.4642492	1.0000000	60.0000000
SBP	56	129.6785714	11.0389511	100.0000000	152.0000000
DBP	56	81.3571429	5.0755334	72.0000000	92.0000000
Age	60	64.0166667	10.3063115	40.0000000	88.0000000

Because you did not include a VAR statement, PROC MEANS summarized every numeric variable in the Blood_Pressure data set (including the subject number). Let's see how to produce a more meaningful report.

Using PROC MEANS Options to Customize the Summary Report

Most of the time, you will want to override the default output and select what statistics you want in your report. The following table shows some of the most popular statistics that you can request using PROC MEANS:

Figure 19.1: Table of PROC MEANS Options

Option	Description
N	The number of nonmissing observations used to compute the statistics
NMISS	The number of missing observations
MEAN	The mean
STD	The standard deviation
CV	The coefficient of variation
CLM	The 95% confidence interval for the mean
STDERR	The standard error
MIN	The minimum value
MAX	The maximum value
MEDIAN	The median
MAXDEC=*n*	The maximum number of decimal places in all the table values

You will probably want to include the MAXDEC= option every time you run PROC MEANS, whether you override the default options or not. This option controls the number of digits to the right of the decimal place for most of the statistics in the output (values for N and NMISS are always integers).

In most cases, you will also want to use the N and NMISS options. These two options add the number of nonmissing and missing values in the output table.

What other options you select will vary depending on what information you need for your particular project. The next program uses MAXDEC, N, and NMISS along with options to print the mean, the standard deviation, the coefficient of variation, the standard error, and the 95% confidence interval for the mean. These statistics are commonly used in reports and journal articles. Here is the program, followed by the output:

Program 19.3: Adding Options to Control PROC MEANS Output

```
title "Running PROC MEANS with Selected Options";
proc means data=Trial.Blood_Pressure n nmiss mean std cv stderr clm
maxdec=3;
   var SBP DBP;
run;
```

A VAR statement was also included in this program to request statistics for the two variables SBP and DBP (systolic and diastolic blood pressure).

> Most of the time, you will want to include a VAR statement when you use PROC MEANS because there may be several numeric variables in your data set for which it makes no sense to compute means and other statistics (such as the subject variable in this data set).

Here is the output:

Output from Program 19.3

Running PROC MEANS with Selected Options

The MEANS Procedure

Variable	N	N Miss	Mean	Std Dev	Coeff of Variation	Std Error	Lower 95% CL for Mean	Upper 95% CL for Mean
SBP	56	4	130.536	10.911	8.359	1.458	127.614	133.458
DBP	56	4	81.321	5.451	6.703	0.728	79.862	82.781

Once you specify any PROC MEANS options (with the exception of MAXDEC=), the output will only contain statistics for the options you specify. Notice that the minimum and maximum values, part of the default set of statistics, are not present in this output.

This summary table displays statistics for SBP and DBP for all subjects in the trial. The next step is to see a breakdown of these statistics by Gender and Drug.

Computing Statistics for Each Value of a BY Variable

It is often useful to compute statistics for each value of some other variable. For example, one of the variables in the Blood_Pressure data set is Gender. You might want to see selected statistics for males and females. One way to do this is to sort the data set by Gender and then include a BY statement with PROC MEANS.

The next program does just this:

Program 19.4: Adding a BY Statement with PROC MEANS

```
title "Statistics for Blood Pressure Study Broken Down by Gender";
proc sort data=Trial.Blood_Pressure out=Blood_Pressure;
   by Gender;
run;

proc means data=Blood_Pressure n nmiss mean std maxdec=2;
   by Gender;
   var SBP DBP;
run;
```

You add an OUT= PROC SORT option so that the data in the original (permanent) data set is not affected. You can use the same data set name (Blood_Pressure) for the output data set in the Work library as you used in the permanent library.

> Most of the time, when you run PROC SORT, you should specify an OUT= procedure option, especially if you are subsetting either observations or variables. This prevents you from damaging the original data set.

Because you added the BY statement to PROC MEANS, you now have your statistics broken down by Gender. Here is a portion of the output:

Output from Program 19.4

Statistics for Blood Pressure Study Broken Down by Gender

The MEANS Procedure

Gender=' '

Variable	N	N Miss	Mean	Std Dev
SBP	2	0	146.00	14.14
DBP	2	0	85.00	1.41

Gender=F

Variable	N	N Miss	Mean	Std Dev
SBP	28	0	129.50	10.90
DBP	28	0	81.43	5.73

Gender=M

Variable	N	N Miss	Mean	Std Dev
SBP	26	4	130.46	10.27
DBP	26	4	80.92	5.34

You now see the N, NMISS, mean, and standard deviation for males and females (as well as the two observations where Gender was missing).

If you want to omit the two observations where Gender is missing, add a WHERE= data set option as part of your PROC SORT, like this:

```
proc sort data=Trial.Blood_Pressure(where=(Gender is not missing))
   out=Blood_Pressure;
```

The output data set, Blood_Pressure, is now sorted by Gender and does not include any observations with missing values for Gender. This also shows why it is usually advantageous to include an OUT= option with PROC SORT. Without it, you would have removed two observations from your permanent Blood_Pressure data set.

Using a CLASS Statement Instead of a BY Statement

You can use a CLASS statement to generate the same information that you produced with a BY statement. One major difference between using a BY statement versus a CLASS statement is that you do not have to sort the data when you use a CLASS statement.

Note: If you have very large data sets and several CLASS variables, there is a possibility that the program will run out of memory, and you will have to use PROC SORT and a BY statement instead of a CLASS statement.

The output you obtain when you use a CLASS statement instead of a BY statement has a slightly different format, but the numbers are the same. The next program shows how to use a CLASS statement:

Program 19.5: Using a CLASS Statement to See Statistics Broken Down by Region

```
title "Using a CLASS Statement with PROC MEANS";
proc means data=Trial.Blood_Pressure n nmiss mean std maxdec=2;
   class Gender;
   var SBP DBP;
run;
```

Notice that this program is using the original permanent data set and no sorting is necessary. Here is the result:

Output from Program 19.5

Using a CLASS Statement with PROC MEANS

The MEANS Procedure

Gender	N Obs	Variable	N	N Miss	Mean	Std Dev
F	28	SBP	28	0	129.50	10.90
		DBP	28	0	81.43	5.73
M	30	SBP	26	4	130.46	10.27
		DBP	26	4	80.92	5.34

Using a CLASS statement produces a slightly more compact and easy-to-read report. Also, it does not include the observations where Gender is missing.

Including Multiple CLASS Variables with PROC MEANS

Because this was a drug study, you will want to see statistics on SBP and DBP broken down by Gender and Drug. This is easily accomplished by listing these two variables on the CLASS statement. Here is the program and the output:

Program 19.6: Using Two CLASS Variables with PROC MEANS

```
title "Using a CLASS Statement with Two CLASS Variables";
proc means data=Trial.Blood_Pressure n nmiss mean std maxdec=2;
   class Gender Drug;
   var SBP DBP;
run;
```

Output from Program 19.6

Using a CLASS Statement with Two CLASS Variables

The MEANS Procedure

Gender	Drug	N Obs	Variable	N	N Miss	Mean	Std Dev
F	Drug A	10	SBP	10	0	128.80	7.73
			DBP	10	0	80.20	4.94
	Drug B	10	SBP	10	0	127.00	15.00
			DBP	10	0	78.00	4.81
	Placebo	8	SBP	8	0	133.50	7.98
			DBP	8	0	87.25	2.82
M	Drug A	10	SBP	9	1	131.78	8.33
			DBP	9	1	82.00	5.10
	Drug B	10	SBP	9	1	124.00	10.25
			DBP	9	1	80.00	6.63
	Placebo	10	SBP	8	2	136.25	9.10
			DBP	8	2	80.75	4.40

You now see blood pressures broken down by Gender and Drug.

Statistics Broken Down Every Way

You can add the PROC MEANS option PRINTALLTYPES to output statistics broken down by every combination of the CLASS variables. To show how this works, here is Program 19.6 rewritten with this option included:

Program 19.7: Adding the PRINTALLTYPES Option to PROC MEANS

```
proc means data=Trial.Blood_Pressure n nmiss mean std maxdec=2
   printalltypes;
   class Gender Drug;
   var SBP DBP;
run;
```

The output now looks like this:

Output from Program 19.7

Using a CLASS Statement with Two CLASS Variables

The MEANS Procedure

N Obs	Variable	N	N Miss	Mean	Std Dev
58	SBP	54	4	129.96	10.51
	DBP	54	4	81.19	5.50

Drug	N Obs	Variable	N	N Miss	Mean	Std Dev
Drug A	20	SBP	19	1	130.21	7.94
		DBP	19	1	81.05	4.96
Drug B	20	SBP	19	1	125.58	12.71
		DBP	19	1	78.95	5.67
Placebo	18	SBP	16	2	134.88	8.39
		DBP	16	2	84.00	4.90

Gender	N Obs	Variable	N	N Miss	Mean	Std Dev
F	28	SBP	28	0	129.50	10.90
		DBP	28	0	81.43	5.73
M	30	SBP	26	4	130.46	10.27
		DBP	26	4	80.92	5.34

Gender	Drug	N Obs	Variable	N	N Miss	Mean	Std Dev
F	Drug A	10	SBP	10	0	128.80	7.73
			DBP	10	0	80.20	4.94
	Drug B	10	SBP	10	0	127.00	15.00
			DBP	10	0	78.00	4.81
	Placebo	8	SBP	8	0	133.50	7.98
			DBP	8	0	87.25	2.82
M	Drug A	10	SBP	9	1	131.78	8.33
			DBP	9	1	82.00	5.10
	Drug B	10	SBP	9	1	124.00	10.25
			DBP	9	1	80.00	6.63
	Placebo	10	SBP	8	2	136.25	9.10
			DBP	8	2	80.75	4.40

You now see statistics for every combination of the CLASS variables. This is a very useful option and you should always consider using it when you have one or more CLASS variables.

Using PROC MEANS to Create a Summary Data Set

Besides producing printed output, you can use PROC MEANS to create a data set containing the same data that was in the printed reports. Let's start out by computing the mean SBP and DBP for the entire data set. To do this, you add an OUTPUT statement. On this statement, you name the output data set and specify what statistics you want in that data set. For this first example, you will name the data set Grand_Mean and request that the mean and the number of nonmissing observations be included in the data set. Here is the program:

Program 19.8: Using PROC MEANS to Create a Data Set Containing the Grand Mean

```
proc means data=Trial.Blood_Pressure noprint;
   var SBP DBP;
   output out=Grand_Mean mean=Grand_SBP Grand_DBP
          n=Nonmiss_SBP Nonmiss_DBP;
run;
```

You want to create a SAS data set with summary information, but you do not want any printed output; hence, you use the PROC MEANS option NOPRINT. This option suppresses any printed output from the procedure.

PROC SUMMARY produces data sets identical to those produced by PROC MEANS. The only difference between the two procedures is that PROC SUMMARY automatically includes a NOPRINT option. So, use PROC MEANS with the NOPRINT option or PROC SUMMARY—your choice.

You use an OUTPUT statement where you specify the name of the output data set (with the keyword OUT=), and you use keywords to specify which statistics you want in the output data set. These keywords are the same ones that you used as PROC MEANS options (see Figure 19.1). You type the keyword, followed by an equal sign, followed by the variable names that you want to use. These variable names are in the same order as the variable names on the VAR statement. The variable Grand_SBP is the mean SBP for all observations in the data set; the variable Grand_DBP is the mean DBP for all observations in the data set.

Even if you are an experienced SAS programmer, it is a good idea to use PROC PRINT to list the contents of the summary data set. Program 19.9 does this:

Program 19.9: Listing of Data Set Grand_Mean

```
title "Listing of Data Set Grand_Mean";
proc print data=Grand_Mean noobs;
run;
```

Here is the listing:

Output from Program 19.9

Listing of Data Set Grand_Mean

TYPE	_FREQ_	Grand_SBP	Grand_DBP	Nonmiss_SBP	Nonmiss_DBP
0	60	130.536	81.3214	56	56

This data set contains the four variables you specified on the OUTPUT statement (two for SBP and two for DBP). SAS has added two additional variables to this data set: _TYPE_ and _FREQ_. The variable _TYPE_ is useful when you have one or more CLASS variables. The variable _FREQ_ is the total number of observations (missing or nonmissing) in the data set. There were 60 observations in data set Blood_Pressure, but because there were four missing values for SBP and DBP, both variables Nonmiss_SBP and Nonmiss_DBP are equal to 56.

Letting PROC MEANS Name the Variables in the Output Data Set

A nice option that you can use with an OUTPUT statement is called AUTONAME. When you include this option, you do not have to name any of the variables in the output data set—PROC MEANS will name them for you. It uses the names on the VAR statement and adds an underscore followed by the statistic you requested. For example, if you want the mean for SBP and DBP, the output data set will use the variable names SBP_Mean and DBP_Mean. Here is Program 19.8 rewritten to use the AUTONAME option:

Program 19.10: Using AUTONAME to Name the Variables in the Output Data Set

```
proc means data=Trial.Blood_Pressure noprint;
   var SBP DBP;
   output out=Grand_Mean mean= n= / autoname;
run;

title "Listing of Data Set GRAND_MEAN";
title2 "Using the AUTONAME Output Option";
proc print data=Grand_Mean noobs;
run;
```

You use the keywords for the desired statistics, followed by an equal sign and no variable names. Because AUTONAME is a statement option, it follows a slash on the OUTPUT statement. Here is a listing of data set Grand_Mean where the AUTONAME option was used:

Output from Program 19.10

Listing of Data Set GRAND_MEAN
Using the AUTONAME Output Option

TYPE	_FREQ_	SBP_Mean	DBP_Mean	SBP_N	DBP_N
0	60	129.679	81.3571	56	56

This author recommends using AUTONAME for two reasons: One, it saves typing; and two, it creates consistent and predictable variable names for all of your statistics.

Creating a Summary Data Set with CLASS Variables

You may want to create a summary data set that contains statistics broken down by one or more variables. One way to do this is to add a CLASS statement to PROC MEANS, along with an OUTPUT statement. This produces a very useful data set. To demonstrate this process, let's output several statistics for the variables SBP and DBP and include both Gender and Drug as CLASS variables. Here is the program:

Program 19.11: Creating a Summary Data Set with Two CLASS Variables

```
proc means data=Trial.Blood_Pressure noprint;
   class Gender Drug;
   var SBP DBP;
   output out=Summary mean= n= std= / autoname;
run;

title "Listing of Data Set SUMMARY";
proc print data=Summary noobs;
run;
```

You are requesting values for the mean, the number of nonmissing values, and the standard deviation for the variables SBP and DBP, broken down by Gender and Drug. Here is a listing of the data set:

Output from Program 19.11

Listing of Data Set SUMMARY

Gender	Drug	_TYPE_	_FREQ_	SBP_Mean	DBP_Mean	SBP_N	DBP_N	SBP_StdDev	DBP_StdDev
		0	58	129.556	81.2593	54	54	11.2244	5.08488
	Drug A	1	20	130.947	80.3158	19	19	10.9822	5.34429
	Drug B	1	20	121.263	79.3684	19	19	6.7728	3.94702
	Placebo	1	18	137.750	84.6250	16	16	9.1761	4.54423
F		2	28	130.857	81.2857	28	28	11.7307	5.55683
M		2	30	128.154	81.2308	26	26	10.7021	4.63299
F	Drug A	3	10	130.600	80.2000	10	10	12.8599	6.42564
F	Drug B	3	10	123.000	79.8000	10	10	5.0990	3.93841
F	Placebo	3	8	141.000	84.5000	8	8	8.8802	5.42481
M	Drug A	3	10	131.333	80.4444	9	9	9.2195	4.21637
M	Drug B	3	10	119.333	78.8889	9	9	8.1240	4.13656
M	Placebo	3	10	134.500	84.7500	8	8	8.7994	3.84522

This may not be what you expected. The first observation (_TYPE_ = 0) is the mean (along with the other requested statistics) for all the observations. The next three observations are statistics broken down by Drug; the next two observations are statistics broken down by Gender; and, finally, the last six observations show the statistics for every combination of Drug and Gender. Notice that the suffix added for the two variables holding the standard deviations is StdDev, not Std (the option you used). The reason is that Std is an abbreviation for the actual term StdDev.

You probably only want the last six observations in this data set. The easiest way to do this is to add the PROC MEANS option NWAY. This option provides your requested statistics broken down by all of your CLASS variables. Here is Program 19.11 rewritten to include the NWAY option:

Program 19.12: Adding the NWAY Option to PROC MEANS

```
proc means data=Trial.Blood_Pressure noprint nway;
   class Gender Drug;
   var SBP DBP;
   output out=Summary mean= n= std= / autoname;
run;

title "Listing of Data Set SUMMARY";
title2 "NWAY Option Added";
proc print data=Summary noobs;
run;
```

Here is the listing of the Summary data set:

Output from Program 19.12

Listing of Data Set SUMMARY
NWAY Option Added

Gender	Drug	_TYPE_	_FREQ_	SBP_Mean	DBP_Mean	SBP_N	DBP_N	SBP_StdDev	DBP_StdDev
F	Drug A	3	10	130.600	80.2000	10	10	12.8599	6.42564
F	Drug B	3	10	123.000	79.8000	10	10	5.0990	3.93841
F	Placebo	3	8	141.000	84.5000	8	8	8.8802	5.42481
M	Drug A	3	10	131.333	80.4444	9	9	9.2195	4.21637
M	Drug B	3	10	119.333	78.8889	9	9	8.1240	4.13656
M	Placebo	3	10	134.500	84.7500	8	8	8.7994	3.84522

You now have the data set you want.

If you sorted the Blood_Pressure data set by Gender and Drug and used a BY statement instead of a CLASS statement, you would not need the NWAY option. The data set would be identical to the one above.

Using a Formatted CLASS Variable

The Blood_Pressure data set also contains the age of each subject in the study. You might want to see the mean systolic and diastolic blood pressures for two or more age groups. Conveniently, if you use a continuous variable such as Age in a CLASS statement and you also include a FORMAT statement, PROC MEANS will use the formatted values of the CLASS variables to compute statistics. Here is an example:

You want to see the mean and standard deviation of SBP and DBP for three age groups:

- Low–50
- 51–70
- 71–High

All you need to do is create a format and include a CLASS and a FORMAT statement when you run PROC MEANS. The next program demonstrates this:

Program 19.13: Using a Formatted CLASS Variable

```
proc format;
   value AgeGroup low-50   = '50 and Lower'
                  51-70    = '51 to 70'
                  71-high  = '71 +';
run;
```

```
title "Using a Formatted CLASS Variable";
proc means data=Trial.Blood_Pressure n nmiss mean std maxdec=2;
   class Age;
   format Age AgeGroup.;
   var SBP DBP;
run;
```

Here is the output:

Output from Program 19.13

Using a Formatted CLASS Variable

The MEANS Procedure

Age	N Obs	Variable	N	N Miss	Mean	Std Dev
50 and Lower	6	SBP	6	0	130.67	7.66
		DBP	6	0	80.33	6.25
51 to 70	40	SBP	38	2	130.16	11.01
		DBP	38	2	81.79	5.24
71 +	14	SBP	12	2	127.67	13.01
		DBP	12	2	80.50	4.10

Using formatted CLASS variables allows you to see all of your statistics broken down by any grouping of continuous variables. It saves time and effort.

Demonstrating PROC UNIVARIATE

One of the most popular procedures for summarizing data, especially for statistical purposes, is PROC UNIVARIATE. This procedure has a lot in common with PROC MEANS. However, you can use statements to produce histograms and probability plots that are not available with PROC MEANS.

To demonstrate this procedure, the next program uses PROC UNIVARIATE to analyze SBP and produce a histogram and Q-Q plot:

Program 19.14: Demonstrating PROC UNIVARIATE

```
title "Demonstrating PROC UNIVARIATE";
proc univariate data=Trial.Blood_Pressure;
   id Subj;
   var SBP;
   histogram;
   qqplot / normal (mu=est sigma=est);
run;
```

If you have a variable such as Subj or ID, be sure to include an ID statement listing that variable. It will be useful in some of the output. You use a VAR statement just as you did with PROC MEANS. Two additional statements, HISTOGRAM and QQPLOT, were added. HISTOGRAM, as the name implies, generates a histogram for all the variables on the VAR statement. *Q-Q plot* is a quantile-quantile plot that statisticians use to help determine deviations from normality. When you request a Q-Q plot, you can add the option NORMAL. This option draws a straight line representing what a normal distribution would look like on the plot. Following this option, you can specify a mean (mu) and a standard deviation (sigma) for your theoretical normal plot. In this example, you want to use the data to estimate these two values. You use the term EST (stands for *estimated*) to request this.

Here is the result:

Output from Program 19.14

Demonstrating PROC UNIVARIATE

The UNIVARIATE Procedure
Variable: SBP

Moments			
N	56	Sum Weights	56
Mean	129.678571	Sum Observations	7262
Std Deviation	11.0389511	Variance	121.858442
Skewness	0.02568868	Kurtosis	-0.2313177
Uncorrected SS	948428	Corrected SS	6702.21429
Coeff Variation	8.51254836	Std Error Mean	1.47514189

This first section shows you the mean, the standard deviation, and several other statistics. For those who are interested, the skewness and kurtosis are values that help determine if the data values are normally distributed. For both of these statistics, values close to 0 indicate a distribution close to normal.

Basic Statistical Measures			
Location		Variability	
Mean	129.6786	Std Deviation	11.03895
Median	128.0000	Variance	121.85844
Mode	120.0000	Range	52.00000
		Interquartile Range	16.00000

Note: The mode displayed is the smallest of 5 modes with a count of 5.

Here you see other measures of central tendency, including the mean and median, along with measures of spread or dispersion:

<table>
<tr><th colspan="5">Tests for Location: Mu0=0</th></tr>
<tr><th>Test</th><th colspan="2">Statistic</th><th colspan="2">p Value</th></tr>
<tr><td>Student's t</td><td>t</td><td>87.90922</td><td>Pr > |t|</td><td><.0001</td></tr>
<tr><td>Sign</td><td>M</td><td>28</td><td>Pr >= |M|</td><td><.0001</td></tr>
<tr><td>Signed Rank</td><td>S</td><td>798</td><td>Pr >= |S|</td><td><.0001</td></tr>
</table>

Here you see statistical tests to test the null hypothesis that the mean is 0:

Quantiles (Definition 5)	
Level	Quantile
100% Max	152
99%	152
95%	150
90%	146
75% Q3	138
50% Median	128
25% Q1	122
10%	118
5%	114
1%	100
0% Min	100

This section of output shows values of your variable at several different quantiles. In this section, you see that the largest value for SBP was 152 and the lowest was 100. The 25th percentile and the 75th percentile (122 and 138, respectively) are also popular measures.

Extreme Observations					
Lowest			Highest		
Value	Subj	Obs	Value	Subj	Obs
100	53	53	146	20	20
112	32	32	148	17	17
114	49	49	150	14	14
114	48	48	150	28	28
114	34	34	152	10	10

The section shows the five lowest and five highest observations in the data set. The section is especially useful to check if you have some extreme values, possibly data errors. You can have PROC UNIVARIATE print out more than five extreme observations by using a procedure option called NEXTROBS=*N*, where *N* is the number of high and low values you want listed in the table. NEXTROBS stands for *Number of EXTReme OBServations*.

Missing Values			
		Percent Of	
Missing Value	Count	All Obs	Missing Obs
.	4	6.67	100.00

Here you see the number of missing values as a count and also as a percentage of all observations (6.67% in this data set).

Demonstrating PROC UNIVARIATE

The UNIVARIATE Procedure

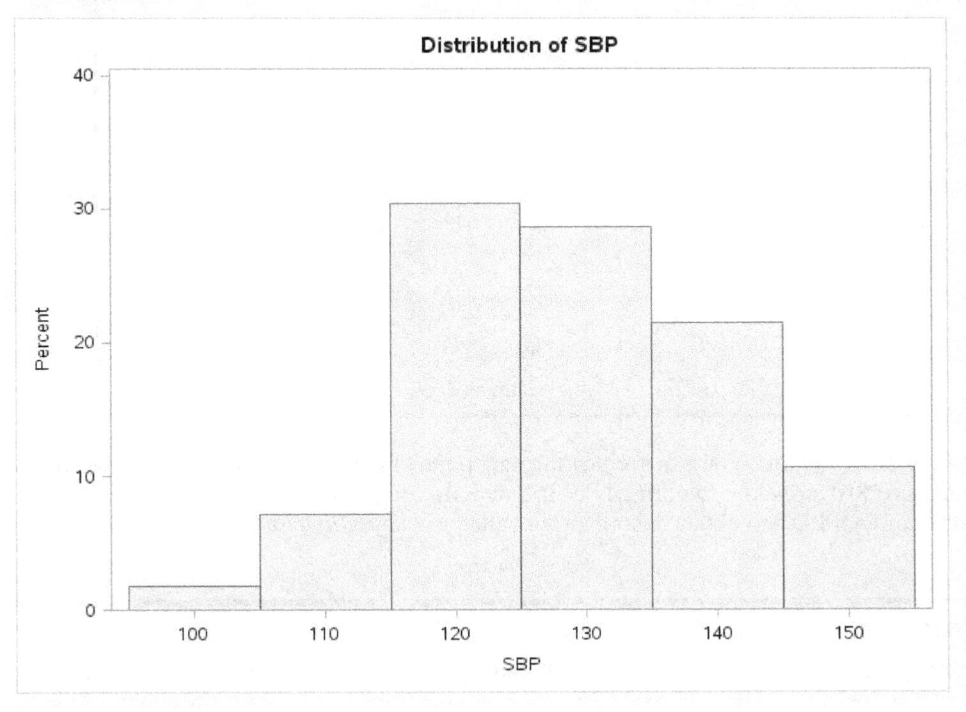

This histogram was produced by the HISTOGRAM statement. If you want to change the bin sizes, you can use a MIDPOINTS= option on the HISTOGRAM statement. For example, to see bars from 100 to 150 but with a bin width of 5 instead of 10, the HISTOGRAM statement would look like this:

```
histogram / midpoints=100 to 150 by 5;
```

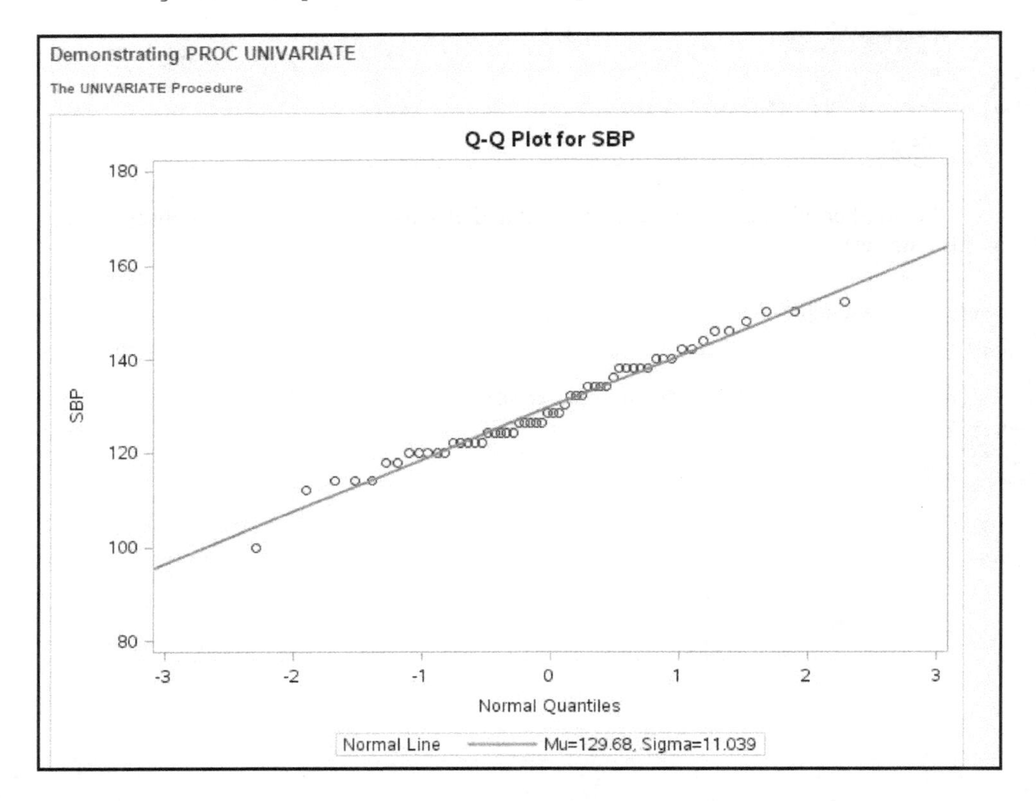

Finally, this is the Q-Q plot. You can see that the data points fall along a straight line, indicating that the values of SBP are close to normally distributed. If you prefer, you can substitute PROBPLOT for QQPLOT to obtain a similar plot called a *probability plot*.

Conclusion

This chapter covered a lot of ground. Besides showing you how to generate summary statistics, you saw how to use PROC MEANS to create a data set containing summary information. This process blurs the distinction between DATA steps and PROC steps. You now know how to create a data set by running a procedure.

There are some topics that were not covered in detail. For example, there are ways to use and interpret the _TYPE_ variable placed in the output data set by PROC MEANS. At some point, you may also want to combine the summary data generated by PROC MEANS with the original raw data. For more information about these topics, I recommend the following book:

Cody, Ron. 2011. *SAS Statistics by Example*. Cary, NC: SAS Institute Inc.

Problems

1. Starting with the Sashelp data set Heart, compute the following statistics for Height and Weight:
    1. Number of nonmissing values
    2. Number of missing values
    3. Mean
    4. Standard deviation
    5. Minimum
    6. Maximum

    Use the appropriate option to print statistics 3 through 6 with two places to the right of the decimal point.
2. Repeat Problem 1 except use a BY statement to compute the statistics broken down by Status.
3. Repeat Problem 2 except use a CLASS statement instead of a BY statement.
4. Repeat Problem 3 except use the two CLASS variables Sex and Status. In addition, use the procedure option to show the statistics broken down by every one of the CLASS variables.
5. Repeat Problem 1 except produce an output data set (Summary) instead of printed output. Use the AUTONAME option on the OUTPUT statement to name all of the variables in the Summary data set.
6. Repeat Problem 5 except produce all of the statistics broken down by Status. Use a CLASS statement and only include observations for each level of Status (i.e., do not include the statistics for all the subjects together).
7. Use PROC UNIVARIATE to analyze the variables Height and Weight from the Sashelp data set Heart. Include statements to produce a histogram for each of these variables.
8. Use PROC MEANS to compute the mean and standard deviation of systolic blood pressure (variable name Systolic) for men only (use a WHERE= data set option to do this) for two groups of men: one group comprising men with weights less than or equal to 150 and the other group for men weighing 151 pounds or more. Do this by using a CLASS statement and writing a format to place men in the two weight groups.

Chapter 20: Computing Frequencies

Introduction

You can use PROC FREQ to create one-way, two-way (row and column), and three-way (page, row, and column) tables. You can output counts and percentages as well as statistics such as chi-square and Fisher's Exact test. This chapter will show you examples of these tasks.

Creating a Data Set to Demonstrate Features of PROC FREQ

The first step is to have some test data to demonstrate features of PROC FREQ. To this end, you can run the program below to generate a data set called Risk that includes the following variables:

Variable	Description
Subj	Subject number
Gender	1=Male, 2=Female
Age	Age in years
Chol	Cholesterol level

Variable	Description
BP_Status	Blood pressure status (High or Low)
Chol_Status	Cholesterol status (High or Low)
Heart_Attack	1-Yes or 2-No

As in the previous chapter, this program randomly generates all of the data. Here is the program:

Program 20.1: Program to Generate Test Data Set Risk

```
proc format;
   value yesno 1 = '1-Yes'
               0 = '2-No';
run;

data Risk;
   call streaminit(12345678);
   length Age_Group $ 7;
   do Subj = 1 to 250;
      do Gender = 'F','M';
         Age = round(rand('uniform')*30 + 50);
         if Age lt 60 then Age_Group = '1:< 60';
         else if Age le 70 then Age_Group = '2:60-70';
         else Age_Group = '3:71+';
         if rand('uniform') lt .3 then BP_Status = 'High';
         else BP_Status = 'Low';
         Chol = rand('normal',200,30) +
rand('uniform')*8*(Gender='M');
         Chol = round(Chol);
         if Chol gt 240 then Chol_Status = 'High';
         else Chol_Status = 'Low';
         Score = .1*Chol + age + 8*(Gender eq 'M') +
         10*(BP_Status = 'High');
         Heart_Attack = (Score gt 100)*(rand('uniform') lt .6);
         output;
       end;
   end;
   keep Subj Gender Age Chol Chol_Status BP_Status Heart_Attack;
   format Heart_Attack yesno.;
run;

title "Listing of Data Set RISK (first 10 observations)";
```

```
proc print data=Risk(obs=10);
   id Subj;
run;
```

Here is a listing of the first 10 observations from data set Risk:

Output from Program 20.1

Listing of Data Set RISK (first 10 observations)

Subj	Gender	Age	BP_Status	Chol	Chol_Status	Heart_Attack
1	F	68	Low	237	Low	2-No
1	M	72	High	210	Low	1-Yes
2	F	50	Low	157	Low	2-No
2	M	71	Low	179	Low	2-No
3	F	64	Low	184	Low	2-No
3	M	71	Low	202	Low	2-No
4	F	76	Low	253	High	2-No
4	M	57	Low	208	Low	2-No
5	F	75	High	207	Low	1-Yes
5	M	57	High	185	Low	2-No

Using PROC FREQ to Generate One-Way Frequency Tables

To create one-way frequency tables, specify the list of variables on a TABLES statement as follows:

Program 20.2: Using PROC FREQ to Generate One-Way Frequency Tables

```
title "One-way Frequency Tables";
proc freq data=Risk;
   tables Gender Heart_Attack;
run;
```

In this example, you are requesting one-way frequencies for Gender and Heart_Attack. PROC FREQ can also compute frequencies for numeric variables. However, beware: If you have a variable such as Age with many different values, PROC FREQ will compute frequencies for every unique value. Here is the output:

Output from Program 20.2

One-way Frequency Tables

The FREQ Procedure

Gender	Frequency	Percent	Cumulative Frequency	Cumulative Percent
F	250	50.00	250	50.00
M	250	50.00	500	100.00

Heart_Attack	Frequency	Percent	Cumulative Frequency	Cumulative Percent
2-No	415	83.00	415	83.00
1-Yes	85	17.00	500	100.00

You see frequency, percent, cumulative frequency, and cumulative percent in the two tables. The order of the values is based, by default, on the internal values of the variables. For numeric variables, lower numbers come before higher numbers; for character variables, the values are sorted alphabetically. Because Heart_Attack was coded as 0 and 1, the first entry in the table is 2-No, the formatted value for the internal value of 0.

You may wonder why the formats '2-No' and '1-Yes' were created. Why not just 'Yes' and 'No'? The answer is that you can request the order in any tables to be based on the formatted value of a variable by using the PROC FREQ option ORDER=formatted. Alphabetically, '1-Yes' comes before '2-No'. That will force the 'Yes' frequencies to be listed before the 'No' frequencies. For the 2 x 2 tables coming up next, it will be preferable (for statistical purposes) to have the 'Yes' values come before the 'No' values.

It is rarely useful to see cumulative frequencies or percentages. You can eliminate these values from the tables with a TABLES option called NOCUM. The program that follows uses both the ORDER= procedure option and the NOCUM statement option:

Program 20.3: Changing the Table Order and Removing the Cumulative Statistics

```
title "One-way Frequency Tables";
proc freq data=Risk order=formatted;
   tables Gender Heart_Attack / nocum;
run;
```

The listing below shows how these two options affect the tables:

Output from Program 20.3

One-way Frequency Tables

The FREQ Procedure

Gender	Frequency	Percent
F	250	50.00
M	250	50.00

Heart_Attack	Frequency	Percent
1-Yes	85	17.00
2-No	415	83.00

The order of the frequencies for the variable Heart_Attack is now controlled by the formatted values for Gender and Heart_Attack, and the cumulative frequencies are no longer included in the tables.

Before we leave this section, you should know one other ordering option: ORDER=freq. This option arranges the frequencies from the most frequent to the least frequent. At times, this can be extremely useful.

Creating Two-Way Frequency Tables

You can create a row by column table by placing one or more variables on the TABLES statement, followed by an asterisk, followed by another list of variables. The variables before the asterisk form the rows of the tables, and the variables after the asterisk form the columns of the tables. If you select more than one variable for the rows or columns list, that list must be placed in parentheses. The syntax is:

```
tables (list of row variables) * (list of column variables) /
options;
```

PROC FREQ will create a table for every combination of variables in the rows list with every variable in the columns list. If you write the following TABLES statement:

```
tables (A B) * (C D E);
```

PROC FREQ will create tables A by C, A by D, A by E, B by C, B by D, and B by E.

To demonstrate a two-way frequency table, let's create a table using the two variables BP_Status and Heart_Attack. Traditionally, epidemiologists like the outcome variable (Heart_Attack, in this example) to form the columns of the table. Here is the program:

Program 20.4: Creating a Two-Way Frequency Table

```
title "Two-way Frequency Table of BP_Status by Heart_Attack";
proc freq data=Risk order=formatted;
   tables BP_Status * Heart_Attack;
run;
```

This table request will produce a table with BP_Status forming the rows of the table and Heart_Attack forming the columns of the table. Here is the listing:

Output from Program 20.4

Two-way Frequency Table of BP_Status by Heart_Attack

The FREQ Procedure

Frequency Percent Row Pct Col Pct

Table of BP_Status by Heart_Attack			
	Heart_Attack		
BP_Status	1-Yes	2-No	Total
High	44 8.80 28.57 51.76	110 22.00 71.43 26.51	154 30.80
Low	41 8.20 11.85 48.24	305 61.00 88.15 73.49	346 69.20
Total	85 17.00	415 83.00	500 100.00

The key to this table is displayed in the upper left-hand part of the output. The top number in each box is a frequency count. For example, there were 44 subjects with high blood pressure who had a heart attack. The second number in the table is a percent: These 44 subjects represent 8.8% of all the subjects in the table (500). The third number in the table is a row percent. Of the 154 subjects who had high blood pressure, 28.57% of them had a heart attack. Finally, the bottom number in the box is a column percentage. Of the 85 subjects who had a heart attack, 51.76% of them had high blood pressure.

The primary numbers of interest to a medical researcher looking at this table would be the percent of people in the high blood pressure group who had a heart attack (28.57%) compared to the percent of people in the low blood pressure group who had a heart attack (11.85%).

One popular statistical test to decide if the difference in these two proportions is statistically significant is called *chi-square*. You can request a chi-square test by adding the CHISQ TABLES option to your program. The modified program is shown next:

Program 20.5: Adding a Request for a Chi-Square Test

```
title "Two-way Frequency Table of BP_Status by Heart_Attack";
proc freq data=Risk order=formatted;
   tables BP_Status * Heart_Attack / chisq;
run;
```

The output contains the same table as above, with the following added information:

Partial Output from Program 20.5

Statistics for Table of BP_Status by Heart_Attack

Statistic	DF	Value	Prob
Chi-Square	1	21.1184	<.0001
Likelihood Ratio Chi-Square	1	19.7866	<.0001
Continuity Adj. Chi-Square	1	19.9500	<.0001
Mantel-Haenszel Chi-Square	1	21.0762	<.0001
Phi Coefficient		0.2055	
Contingency Coefficient		0.2013	
Cramer's V		0.2055	

The CHISQ option produces other statistics that are not shown here. Of primary interest in this table is the first row where you see a chi-square of 21.1184 with a *p*-value of <.0001. This very small *p*-value tells you that if blood pressure status had no relationship to having a heart attack, the probability of getting such a large difference in the proportion between the two blood pressure groups by chance alone is less than .0001. This is considered highly significant by statisticians.

Creating Three-Way Frequency Tables

You can create three-way frequency tables by specifying the page, row, and column variables, separated by an asterisk. You should be cautious when you do this—it may generate a large volume of output if your page variable(s) have large numbers of values. To keep the output small, the example for a three-way table that follows uses Gender as the page variable, BP_Status as the row variable, and Heart_Attack as the column variable. Here is the program:

Program 20.6: Creating a Three-Way Table

```
title "Three-way Table of Gender by BP_Status by Heart_Attack";
proc freq data=Risk order=formatted;
   tables Gender * BP_Status * Heart_Attack;
run;
```

Here is the output:

Output from Program 20.6

Three-way Table of Gender by BP_Status by Heart_Attack

The FREQ Procedure

Frequency Percent Row Pct Col Pct

Table 1 of BP_Status by Heart_Attack			
Controlling for Gender=F			
	Heart_Attack		
BP_Status	1-Yes	2-No	Total
High	18 7.20 23.38 72.00	59 23.60 76.62 26.22	77 30.80
Low	7 2.80 4.05 28.00	166 66.40 95.95 73.78	173 69.20
Total	25 10.00	225 90.00	250 100.00

Frequency Percent Row Pct Col Pct

Table 2 of BP_Status by Heart_Attack			
Controlling for Gender=M			
	Heart_Attack		
BP_Status	1-Yes	2-No	Total
High	26 10.40 33.77 43.33	51 20.40 66.23 26.84	77 30.80
Low	34 13.60 19.65 56.67	139 55.60 80.35 73.16	173 69.20
Total	60 24.00	190 76.00	250 100.00

Inspection of these tables suggests that high blood pressure is related to the incidence of heart attack in females and males. You could verify this by adding the CHISQ option on the TABLES statement.

Using Formats to Create Groups for Numeric Variables

You can use formats to group values to be displayed in the frequency tables. One of the unique features of PROC FREQ is that it automatically uses formatted values for any variable that is associated with a format. As an example, suppose you want to compute frequencies for the variable Age, and you want age groups of 20 years. The program that follows demonstrates how you can use a format to accomplish this task:

Program 20.7: Using Formats to Group a Numeric Variable

```
proc format;
   value Agegroup low-19 = '<20'
                  20-39  = '20 to 39'
                  40-59  = '40 to 59'
                  60-79  = '60 to 79'
                  80-high= '80+'
                     .   = 'Missing';
run;

title "Using a Format to Group a Numeric Variable";
proc freq data=Risk;
   tables Age / nocum;
   format Age Agegroup.;
run;
```

Because you included a FORMAT statement in the procedure, frequencies are computed for the age groups, not the actual age values. Here is the output:

Output from Program 20.7

Using a Format to Group a Numneric Variable

The FREQ Procedure

Age	Frequency	Percent
40 to 59	157	31.40
60 to 79	330	66.00
80+	13	2.60

Frequencies are displayed for age groups. Notice the order of the frequencies. Because the option ORDER= was omitted in the program, PROC FREQ used its default ordering called INTERNAL. So, regardless of the values of the formatted ranges, the order in the table is by increasing age.

Conclusion

PROC FREQ is one of the main procedures for computing frequencies. As you have seen, it can construct one-way, two-way, and three-way tables. In addition, by adding options, you can request a variety of statistical tests to be performed on the resulting tables.

Problems

1. Using the Sashelp data set Heart, compute one-way frequencies and percentages for the variables Status, BP_Status, and Smoking_Status. Use options to omit cumulative frequencies from the resulting tables.
2. Compute frequencies and percentages for Smoking_Status from the Sashelp data set Heart. Use a procedure option so that the table is in frequency order (with the most frequent category listed first).
3. Compute frequencies and percentages for Status from the Sashelp data set Heart. Create formats and use the appropriate PROC FREQ option so that the category "Dead" comes before "Alive" in the resulting table.
4. Using the Sashelp data set Heart, construct a two-way table for the variables Sex (rows of the table) and Status (columns of the table). Arrange the table so that "Dead" comes before "Alive" and "Male" comes before "Female". If you are statistically inclined, add an option to compute the chi-square for this table.
5. Using the Sashelp data set Heart, construct a three-way table with Sex as the page dimension, Weight_Status as the row dimension, and Status as the column dimension. There are three categories for Weight_Status. Write the necessary code to eliminate all observations where Weight_Status is equal to "Underweight".
6. Using the Sashelp data set Heart, construct a two-way table of Systolic (rows) by Diastolic (columns) blood pressures. Both of these variables are numeric. Use formats to divide Systolic and Diastolic into two groups (below 200 vs. 200 and above for Systolic; below 120 vs. 120 and above for Diastolic). If you are interested, include a request for the chi-square.

Appendix: Solutions to the Odd-Numbered Problems

Solution 8-1;

```
data Quick_Survey;
   infile '/folders/myfolders/Quick.txt';
   informat Subj $3.
            Gender $1.
            DOB mmddyy10.
            Income_Group $1.;
    input Subj
         Gender
         DOB
         Height
         Weight
         Income_Group;
   format DOB mmddyy10.;
run;
title "Listing of Data Set QUICK_SURVEY";
proc print data=Quick_Survey;
   id Subj;
run;

*Solution 8-3;
title "Frequencies";
proc freq data=Quick_Survey order=freq;
   tables Gender Income_Group / nocum;
run;
```

Solution 8-5;

```
data Quick_Survey;
   infile '/folders/myfolders/Quick.csv' dsd;
      informat Subj $3.
            Gender $1.
            DOB mmddyy10.
            Income_Group $1.;
    input Subj
         Gender
         DOB
         Height
         Weight
         Income_Group;
```

```
   format DOB mmddyy10.;
run;
title "Listing of Data Set QUICK_SURVEY";
proc print data=Quick_Survey;
   id Subj;
run;
```

Solution 8-7;

```
/** Import an XLSX file.  **/

PROC IMPORT DATAFILE="/folders/myfolders/Grades.xlsx"
          OUT=WORK.MYEXCEL
          DBMS=XLSX
          REPLACE;
RUN;
*Comment removed from the next line;
PROC PRINT DATA=WORK.MYEXCEL; RUN;
```

Solution 8-9;

```
data Formatted;
   infile '/folders/myfolders/Quick_Cols.txt' pad;
   input @1  Subj     $3
         @4  Gender   $1.
         @5  DOB      mmddyy10.
         @15 Height    2.
         @17 Weight    3.
         @20 Income_Group $1.;
    format DOB mmddyy10.;
run;
title "Listing of Data Set Formatted";
proc print data=Formatted noobs;
run;
```

Solution 9-1;

```
title "PROC CONTENTS for SASHELP.HEART";
proc contents data=sashelp.Heart;
run;

title "PROC CONTENTS with the VARNUM option";
proc contents data=sashelp.Heart VARNUM;
run;

*Solution 9-3;
libname easy '/folders/myfolders';
data easy.Heart_Vars;
   set sashelp.heart(keep=BP_Status Chol_Status Systolic Diastolic
Status);
 run;
```

Solution 9-5;

```
libname sasdata '/folders/myshortcuts/sasdata';
data sasdata.Young_Males;
   set sashelp.class(where=(Sex = 'M' and Age in (11 12)));
run;
```

Solution 10-1;

```
proc format;
   value Gender 1='Male' 2='Female';
   value $Ques '1'='Strongly Disagree' '2'='Disagree' '3'='No opinion'
               '4'='Agree' '5'='Strongly Agree';
   value AgeGrp 0-20='Young' 21-40='Still Young' 41-60='Middle'
                61-high='Older';
run;

data Questionnaire;
   informat Gender 1. Q1-Q4 $1. Visit date9.;
   input Gender Q1-Q4 Visit Age;
   format Gender gender. Q1-Q4 $Ques. Visit mmddyy10. Age AgeGrp.;
datalines;
1 3 4 1 2 29May2015 16
1 5 5 4 3 01Sep2015 25
2 2 2 1 3 04Jul2014 45
2 3 3 3 4 07Feb2015 65
;
title "Listing of Data Set QUESTIONNAIRE";
proc print data=Questionnaire noobs;
run;
```

Solution 10-3;

```
proc format;
   value $Grades 'A','B' = 'Good'
                 'C'     = 'Average'
                 'D'     = 'Poor'
                 'F'     = 'Fail'
                 'I'     = 'Incomplete'
                 ' '     = 'Missing'
                 Other   = 'Invalid';
run;
```

Solution 11-1;

```
data Group_Fish;
   set SASHELP.Fish(keep=Species Weight Height);
   if missing(Weight) then Fish_Grp = .;
/* Alternative:
   if Weight = . then Fish_Grp = .;
*/

   else if Weight le 100 then Fish_Grp = 1;
```

```
      else if Weight le 200 then Fish_Grp = 2;
      else if Weight le 500 then Fish_Grp = 3;
      else if Weight le 1000 then Fish_Grp = 4;
      else if Weight ge 1001 then Fish_Grp = 5;
   run;
   title "Listing of first 10 Observations in Group_Fish";
   proc print data=Group_Fish(obs=10) noobs;
   run;
   *Solution 11-3;
   data High_BP;
      set sashelp.Heart(keep=Diastolic Systolic Status);
      if Systolic gt 250 or Diastolic gt 180;
   run;
   title "Listing of High_BP";
   proc print data=High_BP noobs;
   run;
```

Solution 11-5;

```
/*
1. data Weights;
2.    input Wt;
3.    if Wt lt 100 then Wt_Group = 1;
Missing values will be in Wt_Group 1

4.    if Wt lt 200 then Wt_Group = 2;
Should be Else if

5.    if Wt lt 300 then Wt_Group = 3;
Should be Else if
6. datalines;
50
150
250
;
*/
data Weights;
   input Wt;
   if missing(Wt) then Wt_Group = .;
   else if Wt lt 100 then Wt_Group = 1;
   else if Wt lt 200 then Wt_Group = 2;
   else if Wt lt 300 then Wt_Group = 3;
datalines;
50
150
250
;
title "Liting of Weights";
proc print data=Weights noobs;
run;
```

Solution 12-1;

```
data Wt_Convert;
   do Pounds = 0 to 100 by 10;
      Kg = Pounds/2.2;
      output;
   end;
run;

title "Weight Conversion Table";
proc print data=Wt_Convert noobs;
run;
```

Solution 12-3;

```
data Study;
   do Group = 'A','B','C';
      input Score;
      output;
   end;
datalines;
10
11
12
20
21
22
;
title "Listing of Study";
proc print data=Study noobs;
run;
```

Solution 12-5;

```
data Interest;
   Money = 100;
      do until (Money gt 200);
      Year + 1;
      Money = Money + .03*Money;
      output;
   end;
run;

title "Listing of Interest";
proc print data=Interest noobs;
run;
```

Solution 13-1;

```
data Read_Dates;
   input @1 Date1 mmddyy10.
         @12 Date2 date9.;
```

```
   format Date1 Date2 mmddyy10.;
datalines;
10/21/2015 12Jun2015
12/25/2015  9Apr2014
;
title "Listing of Dates";
proc print data=Read_Dates;
run;
```

Solution 13-3;

```
data Dates;
   set sashelp.Retail(keep=Month Day Year);
   SAS_Date = mdy(Month,Day,Year);
   format SAS_Date mmddyy10.;
run;
title "Listing of Dates";
proc print data=Dates(obs=5) noobs;
run;
```

Solution 13-5;

```
data Study;
   call streaminit(13579);
   do Subj = 1 to 10;
      Date = '01Jan2015'd + int(rand('uniform')*300);
      output;
   end;
   format Date date9.;
run;

title "Out of Range Dates";
data _null_;
   set Study;
   where Date lt '01Jan2015'd or Date gt '04Jul2015'd;
   file print; *Send output to Result window;
   put Subj= Date=;
run;
```

Solution 14-1;

```
data Small_Perch;
   set SASHELP.Fish;
   where Species = 'Perch' and Weight lt 50;
run;
title "Listing of Small Perch";
proc print data=Small_Perch noobs;
run;
```

Solution 14-3;

```
data Questionnaire;
   informat Gender 1. Q1-Q4 $1. Visit date9.;
   input Gender Q1-Q4 Visit Age;
   if sum(of Q1-Q3) ge 6;
   format Viit date9.;
datalines;
1 3 4 1 2 29May2015 16
1 5 5 4 3 01Sep2015 25
2 2 2 1 3 04Jul2014 45
2 3 3 3 4 07Feb2015 65
;

title "Listing of Data Set QUESTIONNAIRE";
proc print data=Questionnaire noobs;
run;
```

Solution 14-5;

```
data FirstQtr;
   input Name $ Quantity Cost;
datalines;
Fred 100 3000
Jane 90 4000
April 120 5000
;
data SecondQtr;
   input Name $ Quantity Cost;
datalines;
Ron 200 9000
Jan 210 9500
Steve 177 5400
;
data FirstHalf;
   set FirstQtr SecondQtr;
run;
title "Listing of Data Set FirstHalf";
proc print data=FirstHalf noobs;
run;
```

Solution 14-7;

```
data First;
   input ID $ X Y Z;
datalines;
001 1 2 3
004 3 4 5
002 5 7 8
006 8 9 6
;
data Second;
```

```
   input ID $ Nane $;
datalines;
002 Jim
003 Fred
001 Susan
004 Jane
;
proc sort data=First;
   by ID;
run;
proc sort data=Second;
   by Id;
run;

data Both;
   merge First(in=In_One) Second(in=In_Two);
   by ID;
   if In_One and In_Two;
run;
title "Listing of Data Set Both";
proc print data=Both noobs;
run;
```

Solution 14-9;

```
data Prices;
   input Item_Number $ Price;
datalines;
A123 $123
B76 4.56
X200 400
D88 39.75
;

data New;
   input Item_Number $ Price;
datalines;
X200 410
A123 121
;
proc sort data=Prices;
   by Item_Number;
run;
proc sort data=New;
   by Item_Number;
run;

data New_Prices;
   update Prices New;
   by Item_Number;
run;
```

```
title "Listing of New_Prices";
proc print data=New_Prices noobs;
run;
```

Solution 15-1;

```
data Questionnaire2;
   input Subj $ Q1-Q20;
datalines;
001 1 2 3 4 5 1 2 3 4 5 1 2 3 4 5 1 2 3 4 5
002 . . . . 3 2 3 1 2 3 4 3 2 3 4 3 5 4 4 4
003 1 2 1 2 1 2 12 3 2 3 . . . . . . . 4 5 5 4
004 1 4 3 4 5 . 4 5 4 3 . . 1 1 1 1 1 1 1 1
;

data Score_Quest;
   set Questionnaire2;
   if n(of Q1-Q10) ge 7 then Score1 = mean(of Q1-Q10);
   if nmiss(of Q11-Q20) le 5 then Score2 = median(of Q11-Q20);
   Score3 = max(Q1-Q10);
   Score4 = sum (largest(1,of Q1-Q10), largest(2,of Q1-Q10));
   drop Q1-Q20;
run;

title "Listing of Data Set Score_Quest";
proc print data=Score_Quest noobs;
run;
```

Solution 15-3;

```
data Char_Data;
   length Date $10 Weight Height $ 3;
   input Date Weight Height;
datalines;
10/21/1966 220 72
5/6/2000 110 63
;
data Num_Data;
   set Char_Data(rename=(Date=C_Date Weight=C_Weight
Height=C_Height));
   Date = input(C_date,mmddyy10.);
   Weight=input(C_Weight,12.);
   Height = input(C_Height,12.);
   format Date date9.;
   drop C_:;
   *Note: The colon in the DROP statement says to drop all variables
    that start with C_.  The colon is like a wild-card and says to
    reference all the variables with the same beginning characters;
run;
title "Listing of Data Set Num_Data";
proc print data=Num_Data noobs;
run;
```

Solution 15-5;

```
data Oscar;
   length String $ 10 Name $ 20 Comment $ 25 Address $ 30
          Q1-Q5 $ 1;
   infile datalines dsd dlm=" ";
*Note: the DSD option is needed to strip the quotes from
 the variables that contain blanks;
   input String Name Comment Address Q1-Q5;
   L1 = lengthn(String);
   L2 = lengthc(String);
datalines;
AbC "jane E. MarPle" "Good Bad Bad Good" "25 River Road" y n N Y Y
12345 "Ron Cody" "Good Bad Ugly" "123 First Street" N n n n N
98x "Linda Y. d'amore" "No Comment" "1600 Penn Avenue" Y Y y y y
. "First Middle Last" . "21B Baker St." . . . Y N
;
title "Listing of Selected Variables from Data Set Oscar";
proc print data=Oscar noobs;
   var String L1 L2;
run;
```

Solution 15-7;

```
data Oscar;
   length String $ 10 Name $ 20 Comment $ 25 Address $ 30
          Q1-Q5 $ 1;
   length Two_Three $ 2;
   infile datalines dsd dlm=" ";
*Note: the DSD option is needed to strip the quotes from
 the variables that contain blanks;
   input String Name Comment Address Q1-Q5;
   Two_Three = substrn(String,2,2);
datalines;
AbC "jane E. MarPle" "Good Bad Bad Good" "25 River Road" y n N Y Y
12345 "Ron Cody" "Good Bad Ugly" "123 First Street" N n n n N
98x "Linda Y. d'amore" "No Comment" "1600 Penn Avenue" Y Y y y y
. "First Middle Last" . "21B Baker St." . . . Y N
;

title "Listing of Selected Variables from Oscar";
proc print data=Oscar noobs;
   var String Two_Three;
run;
```

Solution 15-9;

```
Data How_Tall;
   input Ht $ @@;
*Note: the @@ at the end of the INPUT statement allows you
 to place several observations on one line of data;
    Height = input(compress(Ht,,'kd'),12.);
```

```
    if find(Ht,'cm','i') then Height = Height/2.54;
datalines;
65inches 200cm 70In. 220Cm. 72INCHES
;
title "Listing of Data Set How_Tall";
proc print data=How_Tall noobs;
run;
```

Solution 15-11;

```
data Oscar;
   length String $ 10 Name $ 20 Comment $ 25 Address $ 30
          Q1-Q5 $ 1;
   infile datalines dsd dlm=" ";
*Note: the DSD option is needed to strip the quotes from
 the variables that contain blanks;
   input String Name Comment Address Q1-Q5;
   Name = propcase(Name," '");
   Address = tranwrd(Address,'Street','St.');
   Address = tranwrd(Address,'Road','Rd.');
   Address = tranwrd(Address,'Avenue','Ave.');
   Last_Name = scan(Name,-1,' ');
datalines;
AbC "jane E. MarPle" "Good Bad Bad Good" "25 River Road" y n N Y Y
12345 "Ron Cody" "Good Bad Ugly" "123 First Street" N n n n N
98x "Linda Y. d'amore" "No Comment" "1600 Penn Avenue" Y Y y y y
. "First Middle Last" . "21B Baker St." . . . Y N
;
title "Selected Variables from Data Set Oscar";
proc print data=Oscar noobs;
   var Address;
run;
```

Solution 16-1;

```
data Clinic;
   informat Date mmddyy10. Subj $3.;
   input Subj Date Heart_Rate Weight;
   format Date date9.;
datalines;
001 10/1/2015 68 150
003 6/25/2015 75 185
001 12/4/2015 66 148
001 11/5/2015 72 152
002 1/1/2014 75 120
003 4/25/2015 80 200
003 5/25/2015 78 190
003 8/20/2015 70 179
;
proc sort data=Clinic;
   by Subj Date;
run;
```

```
data Diff;
   set Clinic;
   by Subj;
   if first.Subj and last.Subj then delete;
   Diff_HR = Heart_Rate - lag(Heart_Rate);
   *Alternative: Diff_HR = dif(Heart_Rate);
   Diff_Weight = dif(Weight);
   if not first.Subj then output;
run;
title "Listing of Data Set Clinic";
proc print data=Diff noobs;
run;
```

Solution 16-3;

```
* Observation Last_x
      1          .
      2          6
      3          .
      4          7
      5          10;

*Solution 17-1;
data Prob1;
   length Char1-Char5 $ 8;
   input x1-x5 Char1-Char5;
   array x[5] x1-x5;
   array Char[5] Char1-Char5;
   *No need for $ in this array statement because Char1-Char5
    already declared character with a length of 8;
   do i = 1 to 5;
      x[i] = round(x[i]);
      Char[i] = upcase(Char[i]);
   end;
   drop i;
datalines;
1.2 3 4.4 4.9 5 a b c d e
1.01 1.5 1.6 1.7 1.8 frank john mary jane susan
;
title "Listing of Data Set Prob1";
proc print data=Prob1 noobs;
run;
```

Solution 17-3;

```
data Missing;
   input w x y z C1 $ C2 $ C3 $;
   array Allnums[*] x y z;
   array Allchars[*] C1-C3;
   do i = 1 to dim(Allnums);
      if Allnums[i] = 999 then Allnums[i] = .;
   end;
```

```
   do i = 1 to dim(Allchars);
      if find(Allchars[i],'NA','i') then Allchars[i] = ' ';
   end;
   drop i;
datalines;
999 1 999 3 Fred NA Jane
8 999 10 20 Michelle Mike John
11 9 8 7 NA na Peter
;
title "Listing of Data Set Missing";
proc print data=Missing noobs;
run;
```

Solution 18-1;

```
title "Listing of the First 10 Observations in Data Set Fish";
title2 "Prepared by: Ron Cody";
title3 "-----------------------------------------------------";

proc print data=SASHELP.Fish(Obs=10 drop=Length1-Length3);
   id Species;
run;

*Solution 18-3;
proc sort data=SASHELP.Fish out=Fish;
   by Species;
run;

title "Listing of Fish Broken Down by Species";
proc print data=Fish(drop=Length1-Length3);
   by Species;
   id Species;
run;
```

Solution 19-1;

```
title "Statistics for Height and Weight in the Heart Data Set";
proc means data=SASHELP.Heart n nmiss mean std min max maxdec=2;
   var Height Weight;
run;

*Solution 19-3;
title "Statistics for Height and Weight in the Heart Data Set";
proc means data=SASHELP.Heart n nmiss mean std min max maxdec=2;
   class Status;
   var Height Weight;
run;
```

Solution 19-5;

```
proc means data=SASHELP.Heart n nmiss mean std min max maxdec=2
           noprint nway;
   var Height Weight;
   output out=Summary mean= n= nmiss= std= min= max= / autoname;
run;
title "Listing of Data Set Summary";
proc print data=Summary noobs;
run;
```

Solution 19-7;

```
title "PROC UNIVARIATE Statistics for Height and Weight";
proc univariate data=SASHELP.Heart;
   var Height Weight;
   histogram;
run;
```

Solution 20-1;

```
title "Summary Data from SASHELP Heart Data Set";
proc freq data=SASHELP.Heart;
   tables Status BP_Status Smoking_Status / nocum;
run;
```

Solution 20-3;

```
proc format;
   value $Status 'Dead' = '1-Dead'
                 'Alive' = '2-Alive';
run;
title "Summary Data from SASHELP Heart Data Set";
proc freq data=SASHELP.Heart order=formatted;
   format Status $Status.;
   tables Status;
run;
```

Solution 20-5;

```
title "Summary Data from SASHELP Heart Data Set";
proc freq data=SASHELP.Heart(where=(Weight_Status ne 'Underweight'));
   tables Sex*Weight_Status*Status / chisq;
run;
```

Index

E

F

G

H

I

J

L

M

N

O

P